Manual of Repairing & Reconditioning Starter Motors and Alternators

Clinton James

© 2011 Clinton James

The right of Clinton James to be identified as the Author of the Work has been asserted by him in accordance with the Copyright, Designs and Patents Act 1988.

First published in the United Kingdom in 2011 by Swordworks Books

All rights reserved. No part of this publication may be produced, stored in a retrieval system, or transmitted in any form or by any means, without the prior permission in writing of the publisher, nor be otherwise circulated in any form of binding or cover other than in which it is published and without a similar condition including this condition being imposed on the subsequent purchaser.

ISBN 978-1-906512-68-2

Typeset by Swordworks Books
Printed and bound in the UK & US
A catalogue record of this book is available from the British Library

Cover design by Swordworks Books
www.swordworks.co.uk

Manual of Repairing & Reconditioning Starter Motors and Alternators

Clinton James

CONTENTS

FOREWORD	7
INTRODUCTION TO AUTOMOTIVE SYSTEMS	9
GENERATORS AND ALTERNATORS	15
PRIMARY IGNITION SYSTEM	29
SECONDARY IGNITION SYSTEM	33
REPAIRING & SERVICING ALTERNATORS	37
STARTER MOTOR RECONDITIONING	61
STARTER MOTOR TESTING	77
SALE OF FINISHED GOODS	85
VEHICLE BATTERIES	89
GLOSSARY	99

FOREWORD

Repairing alternators and starter motors is often a mystery for many otherwise competent and experienced vehicle repairers, mechanics and do it yourself handymen. Yet there is no reason why it should be. We have tried to demystify the whole process in this manual to enable the layperson to effectively repair and recondition these units, either for their own purposes or to establish a lucrative new business. In times of increasing financial hardship, a reconditioned unit can be very attractive to the motorist, especially if the supplier is able to undercut major commercial companies to make their product even more attractive.

INTRODUCTION TO AUTOMOTIVE SYSTEMS

The basic theory of electricity is simple and easily understood if you are just a little patient and curious.

Electron:
The basic unit of electricity. Think of these little guys as bullets, travelling down the wire. It's the movement of electrons which runs the devices which make our lives, and our cars, so comfortable and convenient.

Voltage:
This is the force (or pressure, if you like) of electricity in the wire. If you think of your garden hose as the wire, the water pressure would be equivalent to the voltage. Older cars run on six volt systems and newer (most 1956 and later) utilise twelve volt systems. All vehicles' manuals specify the system voltage.

Current:
This is the movement of electrons in the wire, expressed in a unit called the Amp. The greater the rate of movement through the wire, the greater the number of amps. Think of this as the speed of the water coming out of the garden hose. When you tighten the nozzle the water shoots out further and faster.

Resistance:
This is a restriction to the movement of electrons through the wire or circuit. The unit of resistance is called the OHM and you can think of it as a kink in that garden hose. The higher the resistance, the more current must flow to overcome it. The more current which flows through an area of high resistance, the hotter the wire will become, ultimately failing. Corrosion, loose terminals and too-small diameter wires are three very, very common causes of resistance.
High resistance is the cause of all electrical failures, with the exception of broken wires and lack of grounding.

Watts:
The unit of power in electricity and the product of Amps x Volts. Why is this important? Because designers of circuits need to know the amount of current required for a given device (such as a fan, horn, light, etc.) in order to figure out which diameter wire to use. Example: a 50-watt brake light, operating on 12 volts, will draw 4.1 amps (4.1 amps x 12 volts = 50 watts). The wire diameter must be large enough to carry the current without heating up and melting off its insulation.

Ground:
All electrical devices must be part of a circuit. That is, electrons must flow from the power source through the device to a ground. In cars, the metal chassis is the ground (that's why the battery's negative lead is bolted to the engine or frame) and the power source is the positive

lead on the battery. Without a ground there is only a potential circuit. No electrons will flow, and therefore nothing will work, unless the circuit ends in a ground.

Note: Some cars and trucks utilised positive ground electrical systems, where the positive lead from the battery connects to the frame and the negative lead goes to the electrical wiring harness. This in no way makes it more difficult to wire or troubleshoot; all that's required is to remember that the system is the reverse of normal systems.

All cars run on Direct Current (DC) electrical systems, as opposed to alternating current (AC) which runs your home. DC is a single wire system. That is, the flow of electricity always runs from the source of current through the device and then to ground. It may do this through any number of connections and through other devices, but tracing the path is straightforward.

For practical purposes, the flow of electricity is now considered to be from positive (voltage, designated by a plus sign +) to negative (ground, designated by a minus sign -). Therefore, your car's battery negative terminal is connected to the metal framework of the car (some older cars, mostly foreign, utilised positive ground systems but this is no longer done).

In order to measure voltage, resistance, direction of current flow and other electrical parameters you need a multimeter. These are devices which have been around for many years and are available at electronics stores and even most home centres. Inexpensive reasonably high-quality meters are all the average hobbyist needs, so don't overspend. All these meters can measure DC, AC, resistance and even small amounts of current. Meters in this price range are fully capable of measuring your car's components accurately, as well as your household system and you can choose either analogue or digital types, depending upon whether you like to read a dial or just a number display. After you purchase one, read the instructions and practice measuring voltages and resistances with it. An hour's practice should make you an expert. When you get accustomed to using a multimeter you will quickly come to appreciate its enormous versatility.

Batteries

A battery is an electrochemical device which converts chemical energy into electrical energy. Cars use lead-acid batteries.

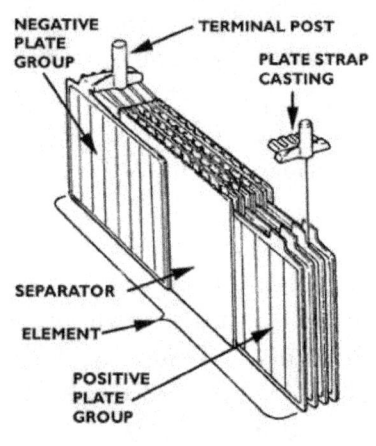

A lead-acid battery uses a series of lead dioxide plates for its positive (+) terminal and porous, soft lead for its negative plates. All the plates are arranged alternately and submerged in a solution of sulphuric acid and water. The positive plate's lead oxide is a compound of lead and oxygen. Sulphuric acid is a compound of hydrogen and the sulphate radical (SO4), so the acid's chemical designation is H2SO4.

Chemically, when a battery is connected to an external load (a device which uses electricity) it begins to discharge. As that happens, the lead in the positive plate combines with the sulphate of the acid, forming lead sulphate (PBSO4) in the positive plate. Oxygen in the positive plate combines with hydrogen from the acid to form water (H2O), which reduces the concentration of the acid in the electrolyte. Also, the pure lead in the negative plate combines with the sulphate, forming lead sulphate and making the positive and negative plates more alike in chemical composition. Electrons are released during this reaction, creating electric current at a specific voltage, 2 volts per cell, with 6 cells in a 12-volt battery.

The battery voltage depends upon the chemical difference between the two plate materials and the concentration of the acid. Because the plates have become more chemically alike and the electrolyte concentration has become weaker, the voltage output gets weaker and weaker until the battery is dead, or discharged.

However, the battery can be re-charged by passing an electrical current through it in the opposite direction of the discharge. The chemical reactions during a charge cycle are the reverse of those that occur during discharge. As the battery is charged the positive plates become lead dioxide again, the negative plates become pure lead again and the electrolyte returns to its proper concentration. The charge-discharge cycle can be repeated over and over again, until fatigue and erosion of the electrodes and corrosion of the positive plates cause eventual failure. Mechanically, batteries are composed of multiple cells, each containing the positive and negative plates. A single cell will produce two volts, so your 12-volt battery has six cells grouped together in one case, for efficiency. The cells are connected in series, or positive to negative to positive to negative; and so on. When you connect something in series you add up each cell's voltage to get the overall battery's output.

So why does such a big, heavy thing like a battery only produce 12 volts? Well, it's the current which does the work and all those plates immersed in that acid are capable of producing impressive amounts of amps, at least for short durations. A typical battery delivers 500-1000 amps and you need all that current to run the starter motor, not to mention other things.

Batteries fail to provide sufficient current generally in only a few ways:
1. The electrolyte and plates wear out. The life of a battery (36 months, 48 months, etc.) is determined by the thickness and

number of plates and you get what you pay for in that regard. Eventually the battery wears out and can't hold a charge. To test for this, have a service station test the cells with a hygrometer (a device which measures specific gravity) or buy one for yourself (they're cheap). If the hygrometer says the battery is shot and it won't hold a charge, replace it.
2. The most common failure of batteries is loose or corroded cable connections. In either case, the reason for failure is high resistance! (Remember, a poor mechanical connection means that little or no current can pass through). If the cables are loose, tighten them thoroughly. If corroded, remove them and clean them with a file or sandpaper (clean both the cable connectors and the terminals!) It's a good practice to clean the connections at least once a year.
3. Overcharging, either through external chargers or faulty regulator, kills batteries by creating so much heat (due to current flow) that the water in the electrolyte is boiled off. In some cases the battery explodes. Of course, connecting jumper cables incorrectly can result in a dead short, with catastrophic consequences (A dead short is when all the current from the voltage source is connected directly to ground without passing through any device or resistance. In the case of a battery, it would be equivalent to connecting both terminals together, causing a huge current flow through the plates, in turn causing massive heating, then boiling, and finally the battery will blow up).

GENERATORS AND ALTERNATORS

How that battery gets charged. In older cars (before about 1964) this was done with a generator. After that all cars switched to alternators and the reasons for the change will become clear. This is how each works.

The Generator:
The basic principle at work here is that electricity produces magnetism. Conversely, magnetism produces electricity. If a current-carrying coil of wire is placed around a bar of steel, the bar will become magnetised. The more turns of wire and the stronger the current, the more powerful the magnet. By placing a soft iron core within the coil, the magnetic force lines are concentrated and strengthened. As there is less electrical resistance in the iron than in the surrounding air, the force lines will follow the core.

The two pole shoes of a generator are constructed in this way. Rather than use magnets, which are heavy and expensive, many turns of wire are wound around the pole shoes. When a current passes through these windings the pole shoes become electromagnets, called Field Coils. These two field coils are connected in series (current passes through one and then through the other) and wound so that one becomes the north pole and the other the south pole of the magnetic field.

STARTER MOTORS & ALTERNATORS

Inside the generator is a spinning central shaft which is supported in bearings at each end. Loops of wire (armature windings) are wound on a special laminated holder called the Armature. The armature is turned by placing a pulley on one end of the shaft and driving it with a V-belt from the engine's crankshaft.

Attached to the armature are electrical contact segments, called the Commutator. These segments are electrically insulated from the armature, and each other, but each is soldered to one of the armature windings. It is the commutator which distributes electricity to the armature in an on-off manner, creating a magnetic field around the armature. Riding over the spinning commutator segments are carbon brushes. These brushes are held in spring-loaded brackets and that pressure holds them against the commutator. It is the brushes which wear out over time and require replacement.

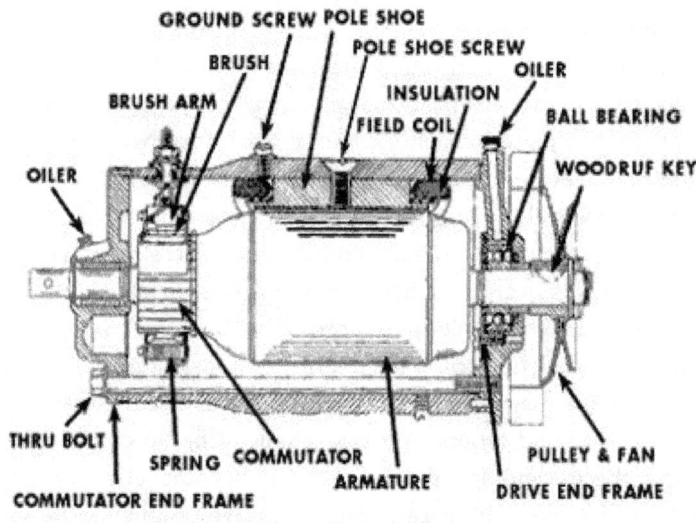

TYPICAL GENERATOR

When the generator armature first begins to spin, there is a weak residual magnetic field in the iron pole shoes. As the armature spins, it begins to build voltage. Some of this voltage is impressed on the field windings through the generator regulator (commonly called the Voltage Regulator). This impressed voltage builds up a stronger winding current, increasing the strength of the magnetic field. The increased field produces more voltage in the armature. This, in turn, builds more current in the field windings, with a resultant higher armature voltage. This voltage could, of course, continue to increase indefinitely, but it is limited (by regulation) to a pre-set peak. At this point all this sounds like perpetual motion, doesn't it? Remember, though, that the energy driving all this is the engine's crankshaft!

It should be noted that the most common failure of a generator is the brushes. Second to that is bearing failure, especially the bearing next to the drive pulley (improper belt tension hastens the demise of this bearing).

A major failure-mechanism in generators is improper installation of a new or rebuilt one. Mechanically the installation is straightforward but electrically, things are more complex. When the generator stopped the last time, residual magnetism remained in the pole shoes. The polarity of the shoes at that time depended on the direction of current flow in the field coil windings. If during testing and rebuilding current is caused to flow in the opposite direction, the pole shoes will change polarity. If the generator is then run in

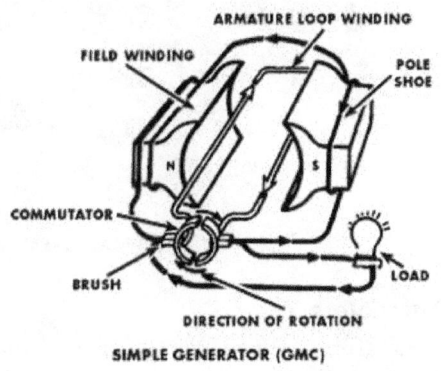

SIMPLE GENERATOR (GMC)

the car, the reversed polarity will cause current to flow in the wrong direction, damaging the regulator and discharging the battery when the car is left overnight. Therefore, all generators must be polarised after installation and before running the car. This is done by holding one end of a wire on the battery terminal of the regulator and scratching the other end against the generator's output terminal (for externally-grounded generators). For internally-grounded generators the proper way to polarise is to disconnect the field lead from the regulator and scratch it on the battery terminal on the regulator.

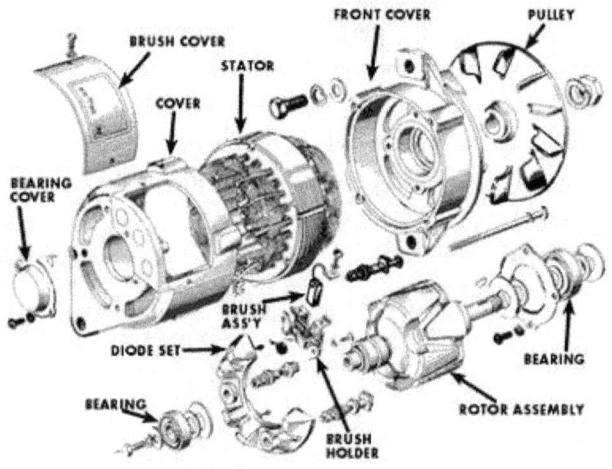

TYPICAL ALTERNATOR

Alternators

Generators produce Direct Current. Alternators produce Alternating Current, AC. Alternators have the advantage of producing far more current at low speeds than generators, thus allowing more and more accessories in the car. In an alternator, the field windings are placed around the spinning central shaft rather than on shoes as in the generator. Two iron pole pieces, cast with fingers, are slid on the shaft, covering the field winding so that the fingers are interspaced.

The fingers on one pole piece form the North poles and the fingers on the other form the South poles. This assembly is called the Rotor. Surrounding the rotor are a series of windings around laminated iron rings, attached to the alternator's case. This assembly is called the Stator. The engine's crankshaft spins the rotor.

Direct current from the battery is fed through into the rotor's field coil by using brushes rubbing against slip-rings. One end of the field coil is fastened to the insulated brush, while the other end is attached to the grounded brush. As the pole fields pass through the stator, current is electromagnetically produced (as in the generator) but since the rotor is composed of alternating North and South poles the current produced flows in opposite direction every 180-degrees of rotation. In other words, the current is alternating.

Why is this more efficient? Well, the stator windings are made up of three separate windings. This produces what is known as three-phase AC. When only one winding is used, single-phase current results (like in a generator). In effect, the alternator produces three times the current of a generator for the same effort on the engine's part. Also, alternators are considerably lighter and smaller than generators.

There's a small problem with alternators, though. AC electricity doesn't work in a car! The car's electrical system, and battery, need DC. Therefore, the alternator's output is rectified into DC. This is done by passing the AC into silicon diodes. Diodes have a peculiar ability to allow current to flow readily in one direction only, stopping the flow if the direction reverses. Multiple diodes are arranged in alternators so that current will flow from the alternator to the battery (in one direction only, creating DC) but not from the battery to the alternator.

In actual operation, the voltage regulator senses the battery voltage and overall demand on the car's electrical system. When charging is needed, the regulator applies battery voltage to the stator's brushes and this creates the electrical field for charging. As the system's demand for charging decreases the voltage to the brushes cuts off. All of this occurs many times per minute, with the system turning on and off repeatedly to keep everything at optimum operating efficiency.

Voltage Regulators

There are no means of internally controlling the output of generators. In other words, the faster it spins the more voltage goes into the car's electrical system. If this was not controlled the generator would damage the battery and burn out the car's lights. Also, if the generator wasn't cut out from the car's circuitry when not running, the battery would discharge through its case.

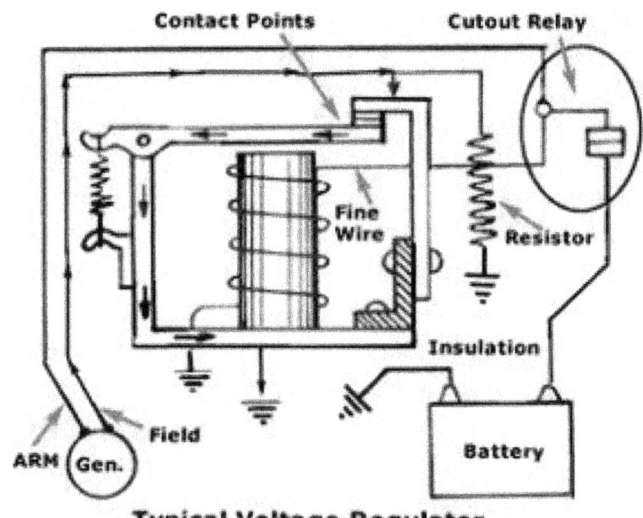

Typical Voltage Regulator

That's where the Regulator (commonly called the Voltage Regulator, but that's only one component of the system) comes in.

Regulators have seen many design improvements over the years, but the most commonly used electro-mechanical regulator is the three-control units in one box type. Let's look at how these things work.

Cutout Relay

Sometimes called the circuit breaker, this device is a magnetic make-and-break switch. It connects the generator to the battery (and therefore the rest of the car) circuit when the generator's voltage builds up to the desired value. It disconnects the generator when it slows down or stops.

The relay has an iron core that is magnetised to pull down a hinged armature. When the armature is pulled down a set of contact points closes and the circuit is completed. When the magnetic field is broken (like when the generator slows down or stops) a spring pulls the armature up, breaking the contact points.

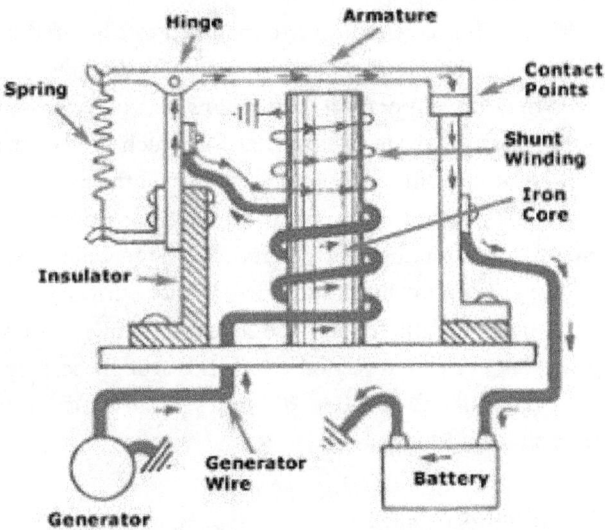

A Schematic of the cutout relay. Trace the flow of current through its components.

An obvious failure mode is the contact points. As they open and close, a slight spark is generated, eventually eroding the material on the points until they either weld themselves together or become so high in resistance that they won't conduct current when closed. In the first case the battery would discharge through the generator overnight and in the second there would be no charging to the system.

Voltage Regulator

Another iron core-operated set of contact points is used to regulate maximum and minimum voltage at all times. This circuit also has a shunt circuit (a shunt re-directs electrical flow) going to ground through a resistor and placed just ahead (electrically) of the points. When the points are closed the field circuit takes the easy route to ground but when the points are open the field circuit must pass through the resistor to get to ground.

The field coil on the generator is connected to one of the voltage regulator contact points. The other point leads directly to ground. When the generator is operating (battery low or a number of devices running) its voltage may stay below that for which the control is set. Since the flow of current will be too weak to pull the armature down the generator field will go to ground through the points. However, if the system is fully charged the generator voltage will increase until it reaches the maximum limit and current flow through the shunt coil will be high enough to pull the armature down and separate the points. This cycle is repeated over and over in real time. The points open and close about 50 to 200 times per second, maintaining a constant voltage in the system.

Current Regulator

Even though the generator's voltage is controlled it is possible for its current to run too high. This would overheat the generator, so a current regulator is incorporated to prevent premature failure.

Similar in appearance to the voltage regulator's iron core, the current regulator's core is wound with a few turns of heavy wire and connected in series with the generator's armature.

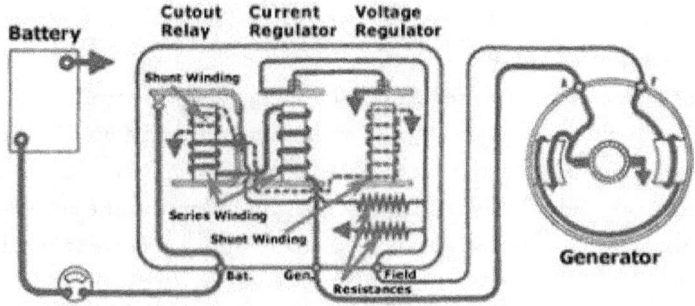

Diagram of a typical three-unit regulator. Trace all circuits a few times and you will understand how it functions.

In operation, current flow increases to the predetermined setting of the unit. At this time, current flow through the heavy wire windings will cause the core to draw the armature down, opening the current regulator points. In order to complete the circuit the field circuit must pass through a resistor. This lowers current output, points close, output increases, points open, output down, points close, and so on. The points, therefore, vibrate open and closed much as the voltage regulator's points do, many times every second.

Because they are mechanical, voltage regulators are easy to troubleshoot. If you study the function of each of the three parts and how they interrelate, it becomes obvious which part is malfunctioning, depending upon symptoms. That means anyone who understands how everything works can easily troubleshoot problems.

Point gaps and spring pressures determine the voltage/current limits and they are exceedingly hard to adjust. Sometimes it can be done on the car using a voltmeter, but generally it is best to replace the entire regulator assembly when a certain part of it fails. Factory assembly of regulators required relatively sophisticated measurement instruments. Adjusting them by feel is a matter of luck, and frequently can result in damage.

Alternator Regulators

The same type of regulator was originally incorporated into alternator-fitted cars and they work in much the same way. However, since some cars used ammeters no current regulator was needed. Therefore, a single unit regulator was used to turn on the alternator's stator windings. It was just a regulator without a current regulator section.

It wasn't long thereafter that the automobile companies converted to transistor voltage regulators. Utilising Zener diodes, transistors, resistors, a capacitor and a thermistor, these regulators maintain proper voltage and current flow throughout the system. Their circuitry operates as fast as 2,000 times per second and they are tremendously reliable. On the other hand, these regulators aren't easy to repair. They are designed to be thrown away and replaced.

Many solid state regulators are mounted inside the alternator and are not serviceable other than the ability to set the voltage limits. That's fine, because they work very well for long periods of time. To check their operation, just measure the battery voltage while the engine is off, then when it's running. You should see something between 13 and 15 volts when running. No change in voltage means either the regulator or alternator isn't working, while higher voltage means the regulator isn't properly regulating.

By now you are becoming more familiar with how electricity is stored, generated and regulated in the automobile. It might have occurred to you that the electrical system only needs 80 to 100 amps of current for general running, even when all accessories are operating.

The main reason for the battery's storage capacity is to operate the starter, and a quick look at the numbers will demonstrate why this is so important:

> *For example, a 500 amp rated battery.*
> *At 12 volts, this 500 amp battery is capable of putting out 6000 watts (amps X volts = watts). We need all the wattage we can get to develop enough horsepower to turn the engine over for ignition and one horsepower (or the power necessary to lift 550 pounds one foot in one second) equals 746 watts. Our battery, therefore, puts out just over eight horsepower. That's just enough for a couple hundred revolutions of the engine before the charge is exhausted.*

Starters are incredibly strong motors that work in a hostile environment. They are the most important part of the starting system, or circuit, consisting of the following:

Flywheel Ring Gear
This is a toothed ring that is fitted to the outside of the engine's flywheel. Matching teeth on the starter motor mesh with this gear in order to spin the crankshaft.

Starter Solenoid (Relay)
All cars are wired so that the battery's main cable connects to the starter motor windings (the thick cable is needed for large current flow). This wire must be switched on and off. It would be costly and inefficient to route it through the ignition switch (not to mention the size of the switch's components required to carry such current). Consequently, a relay is necessary.

Relays are devices that utilise a central iron core fitted closely to the inside of a coil of wire. When the wire is energised the iron core will be drawn down the length of the coil, the direction dependent upon the direction of current flow. If the relay's iron core is fitted with large, high current-carrying contacts it can be used as a high-current switch. Relays are used throughout cars (for horns, electric fans, air conditioning clutches, etc.) and the most important one is the starter solenoid.

The starter solenoid has very large contacts to carry the battery's full current. Its wire coil is actuated by a smaller current from the ignition switch, at which time the iron core slams down to make contact and turn on the starter motor. Most non-Ford starter motors employ a solenoid built into the motor itself. This type of solenoid not only provides the motor's electrical power but also mechanically engages the starter's drive gear onto the flywheel. It is commonly known as the Bendix type of solenoid.

Such solenoids operate in three stages, the disengaged, partially engaged and engaged. In the disengaged position the drive gear is released and no current is flowing. In the partially engaged stage, current from the starter switch flows through both the pull-in and the hold-in coils. Both coils draw the plunger inward, causing it to pull the shift lever and engage the pinion gear. When the plunger is pulled into the coil all the way, the pinion fully engages the ring gear. When the ring gear is fully engaged, engine cranking begins. When the engine starts the hold-in coil will cut out and the plunger will move out, retracting the pinion and opening the starter switch.

Starter Motor

This is a powerful electric motor that engages the car's flywheel in order to spin the crankshaft. As in all electric motors, the starter is composed of windings of wire that form loops, ending at the commutator segments. The armature coils are mounted on the motor's central shaft (supported with bearings) and the field coils are formed into four or more shoes, placed inside the steel frame of the starter. Brushes are used to create electrical contact to the commutator segments and when current is fed into two of the four brushes, it flows through all the loops of the armature and shoe windings and out the other two brushes. This creates a magnetic field around each loop. As the armature turns, the loop will move to a position where the current flow reverses. This constant reversal of current flow allows the armature and field coils to repel each other and spin the motor. The greater the current flowing in the coils, the greater the magnetic forces, and the greater the power of the motor.

The copper loops and field windings are heavy enough to carry a large amount of current with minimum resistance. Since they draw heavy amounts of current, they must not be operated on a continuous basis for longer than 30 seconds. After cranking for 30 seconds it is wise to wait a couple of minutes to let the starter motor dissipate some of its heat. Starters heat quickly, so prolonged use can cause serious damage. A typical symptom of overheating starter motors is extremely slow, laboured engine-cranking.

Various wiring designs are used in starter motors and one of the most popular is the four pole, three winding setup. Two of the windings are in series with themselves and the armature. One winding does not pass through the armature, but goes directly to the ground. This Shunt Winding aids with additional starting torque. However, as the starter speed increases, the shunt still draws a heavy current and tends to keep starter speed within acceptable limits.

Starter motors fail mostly due to overheating. They are placed in a hostile, hot environment and cannot be expected to last indefinitely. Another mode of failure is a shorted or open winding. This exhibits itself as a dead spot on the commutator. If a brush lands on a dead spot the motor won't turn at all.

A third failure-mode is a faulty pinion engagement. Sometimes the pinion assembly gets stiff or stuck due to lack of lubrication or wear. Starter motor rebuilding or replacement is required for all of these problems.

Before doing so, however, check to make sure the electrical connections on the starter, and the battery, are clean and tight. Most failures to start are due to loose or corroded battery cable connections, or low-current solenoid connections, and not faulty starters.

So far we have covered basic DC electricity theory, then batteries, generators, alternators and starters. Adding those together, we have developed an electrical circuit to get the car started and charge the battery, but now we need a circuit to actually run the car. That, of course, is the ignition system and it is comprised of several parts, comprising two distinct systems, the primary and secondary ignition systems.

PRIMARY IGNITION SYSTEM

The primary system consists of the ignition switch, coil primary windings, distributor contact points, condenser, ignition resistor, and starter relay.

Ignition Switch
The ignition switch does at least three things. First, it turns on the car's electrical system so that all accessories can be operated. It does so by providing power to the fuse panel (for those components that are controlled by the switch. Some items are independent of the ignition switch, such as headlights, horn, clock, etc.) When you insert the key and turn the switch to the accessories position, you are turning on the other devices in the car, such as the radio, heater, power windows, seats, defroster, etc.

Second, in the run position, everything is turned on, plus the engine's electrical components that enable it to run. Most important, it turns on the entire primary ignition system.

Third, when you move the switch to the start position it energises the starter.

The ignition switch doesn't carry the necessary current to the starter. It sends a small current to a special device called a Relay that, in turn, allows the starter to crank.

STARTER MOTORS & ALTERNATORS

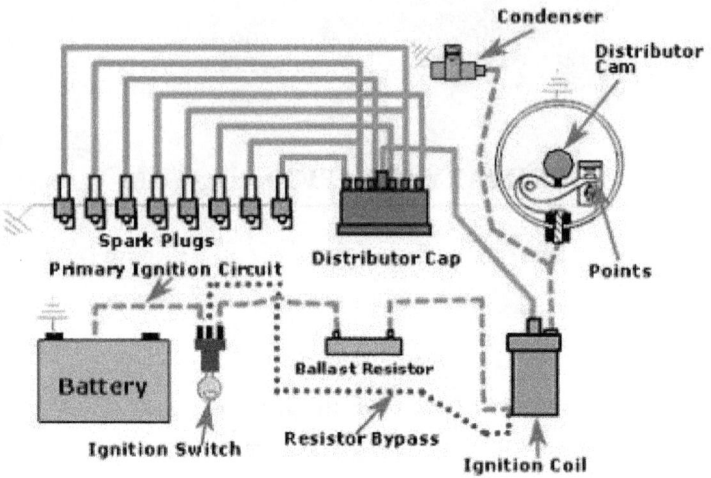

Schematic of Ignition System

The Coil's Primary Winding

Inside the coil are two sets of wound wire, comprising of the primary and secondary windings. The primary windings carry battery voltage through and create a large magnetic field inside the coil. Although the coil's primary windings receive voltage from the ignition switch, they are actually turned on and off by the distributor contact points.

The Distributor Contact Points

These points are opened and closed by a cam on the distributor's main shaft. As it spins the cam's lobes move the actuator outward, disengaging the contacts. When the lobe passes, the contacts close, turning on the coil primary windings. The amount of time the points remain closed is referred to as dwell, and is an important factor in engine tuning.

Condenser

Attached to the points is a condenser, an electrical device (capacitor) that limits current flow through the points to increase their life. The condenser is necessary because the points are opening and closing rapidly, and as they do so the voltage/current is interrupted. This causes an arc, or spark, between the contact points. Over time, this arcing will erode the material on the points and deposit carbon, and eventually the points will not pass current. The condenser acts as a current-absorber to limit the amount of arcing as the points open and close.

Ignition Resistor

The next component is the ignition resistor. It is necessary because ignition coils are designed to step up battery voltage high enough, and fast enough, to keep the engine running at high rpm. That means that, as designed, the coil would produce too much high voltage at low rpm and heat up. Automakers long ago realised that there were two solutions to the problem: using two coils (one for low rpm and one for high) or an ignition resistor. Obviously, the resistor approach is the least expensive and most reliable, so that's what they did. The resistor used varies is resistance as a function of temperature, and limits the voltage to the coil accordingly. As the engine revs up the resistance lowers, allowing more voltage to the coil for fast running, and the reverse happens when the engine slows down. At idle, for instance, only about 7 volts is going through the coil primary windings. The only time the resistor is out of the circuit is during start-up, when the engine needs all the spark it can get. It's bypassed in the ignition switch's start position so that, during starting, the coil gets full battery voltage. Ignition resistors can take many forms, depending upon the manufacturer of the vehicle. Some builders mounted a big resistor on the firewall and some others used a special type of wire (resistance wire) running from the ignition switch to the coil. Still others used coils that were built with an internal resistor. None of these is any better an approach than the others, but it's important to know which type you have.

SECONDARY IGNITION SYSTEM

The secondary ignition system consists of the coil secondary windings, distributor cap, rotor, plug wires and spark plugs.

Coil Secondary winding
It works by induction. If you put a certain voltage through a wire (the primary) that has another wire wrapped around it, the second (hence, secondary) wire will receive an induced voltage from the first. The induced voltage is a function of the number of turns of wire wrapped around, so if you have two coils wrapped around the wire you'll get twice the voltage, and so on. Voltage can be stepped-up and stepped-down using inductance. Transformers are inductance devices, so a coil is a transformer.

Automotive coils generally have secondary-to-primary ratios of 200 to 1. Therefore, a 12-volt input to a coil's primary windings will result in a 24,000-volt output from the secondary winding. That's where the spark plugs get their electricity.

Inductance isn't perpetual motion, nor is it free energy. There are many other considerations to be concerned about. The main one is the coil's inability to hold the induced voltage once it's been built up. In a very short time the voltage bleeds off, leading to a weak spark. Also, the coil takes a finite amount of time to build the charge up. That's the dwell time, normally defined as the degrees of rotation of the camshaft during which the points are closed. Too little dwell and the coil doesn't have time to charge up fully. Too much dwell and the coil has bled off some charge, causing a weak spark. Hesitation, low power, misfiring, pinging and a number of other conditions are symptoms of incorrect dwell.

Distributor Cap and Rotor
The distributor cap is one of the most appropriately-named devices on the car. Its job is to distribute the high voltage pulses, generated by the coil, to the correct spark plug at the correct time. It does this through the rotor. The rotor is keyed to the distributor shaft. On the rotor is a spring-loaded wiper arm, whose purpose is to pick up high voltage pulses from the coil. The wiper arm is electrically connected to the rotor's tip.

Inside the distributor cap are metal nipples that are attached to the sockets holding the plug wires. As the rotor moves around its tip comes within about one millimetre of the cap's nipples, whereupon the high voltage charge jumps over. From there it travels through the plug wire to the plug, which is grounded to the engine block. The charge has nowhere else to go but to the plug's electrode, creating a spark.

Plug Wires
Spark plug wires are greatly underappreciated and often overlooked when it comes to maintenance. They are designed to carry 20,000 to 40,000 volts (much more in modern cars) to the sparkplugs without losing the charge, breaking down electrically or leaking to ground. They operate in extreme heat and vibration environments.

Plug wires were originally constructed with a central copper conductor, wrapped in various layers of insulation. This was very effective, but when AM radios appeared they caused interference (high voltage creates large electromagnetic fields, that in turn cause spurious radio signals. These are picked up by radio sets as static). By the 1950s many manufacturers turned to resistance wires to cure the interference problem.

Resistance wires utilise a central core made up of some flexible material impregnated with a conducting medium, usually a form of carbon, wrapped in insulation. These wires have specific internal resistance that is designed to provide proper spark with minimal electromagnetic interference. Such wires are easily damaged, especially at the ends, where the inner cores are connected to metal boots.

Resistance wires have a finite lifetime and must be replaced after a specific number of miles or operating hours. Solid-core wires also must be replaced when the insulation becomes cracked or stiff.

Spark Plugs

A typical spark plug has various parts. The terminal is the top of the plug where the wire connects. Under that is a ceramic (insulator) section with ribs moulded onto it to reduce flashover. Under that is the crimp, where the metal body begins. Below the crimp are the wrench flats, a hexagonal area that is sized for a specific wrench. Under that is the shell, which is threaded at the bottom to the size (diameter) and reach (depth) of the threaded hole in the engine's cylinder head. The plug ends at the bottom, where there is a ground strap or other device protruding over the central metal core, the electrode. Surrounding the electrode is ceramic insulation to keep it from sparking into the inside of the metal shell rather than at the end ground.

Spark plugs are designed with specific heat ranges. That is, the amount of the central insulator/electrode exposed to the heat of combustion. The deeper the electrode/insulator, and the ground piece, extend into the combustion chamber, the hotter the plug and the less it extends, the colder the plug. Manufacturers specify certain heat ranges for certain conditions, even within the same engine designs.

Plugs also come in Types. Type indicates whether the plug's core is a resistance-type (similar in design to resistance wires) or solid-metal core, projected core nose, and single or multiple ground electrode. Depending on the engine design, plugs may be specified to require a metal gasket between the shell and the threads.

Spark plugs vary tremendously. Plug manufacturers publish extensive applications manuals that clearly spell out design differences and, most importantly, which plugs will work efficiently in the engine application.

REPAIRING & SERVICING ALTERNATORS

The repair and reconditioning of alternators can be a highly lucrative business. However, the basic knowledge behind it has never been readily available without either serving an apprenticeship with an auto electrician, or undertaking extensive research sifting through technical literature. That is why, to date, this field has been confined primarily to auto electricians and a handful of specialist companies. This situation exists not because it requires advanced skills or sophisticated equipment, or that it is an expensive operation to set up, or that it is difficult or expensive to build up a stock, stocks are readily available at minimal cost.

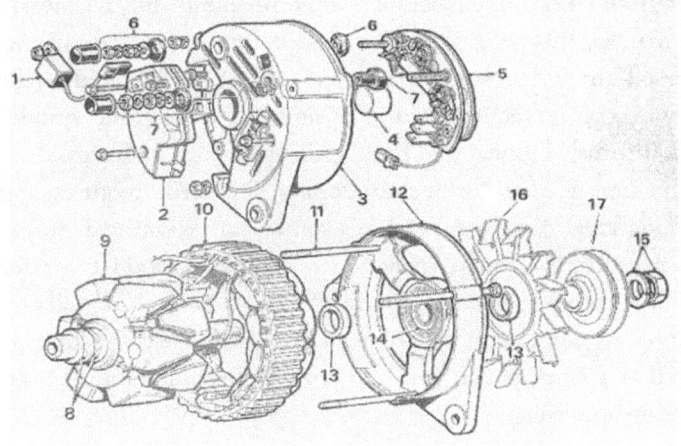

Alternator exploded diagram

1. Suppressor
2. Regulator and brushes
3. Casing / slip ring end bracket
4. Slip ring end bearing
5. Rectifier pack
6. Phase terminal (rev counter feed on diesel)
7. Main terminal
8. Slip rings
9. Rotor
10. Stator
11. Through bolts
12. Drive end bracket
13. Spacers
14. Drive end bearing
15. Pulley nut and washers
16. Cooling fan
17. Pulley

This situation exists because some people in the business, who are making the profits, have a vested interest in minimising the competition and as such do not readily make available the relevant knowledge. In fact the entire business is really quite simple and straightforward once the basic principles are understood. The equipment needed for reconditioning alternator requires a good working area. A garage, shed or even a spare room will suffice but one that offers adequate space to store finished products, preferably on some type of shelving system for ease of access. As well as good lighting to work with there needs to be adequate heating. Being seated at a work bench for long periods in cold conditions can be quite demoralising.

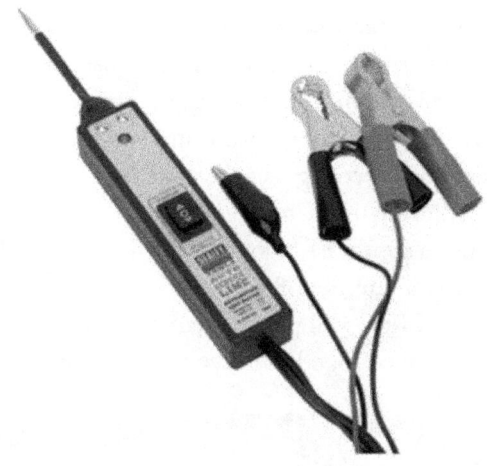

You will also require:

- a small bearing puller
- a main manufacturer's alternator catalogue
- a main manufacturer's cross reference catalogue
- a spare parts manual
- a rotor/stator tester
- a variable range circuit tester or multimeter
- an engineer's vice
- an adjustable bench lamp
- a soldering iron, solder
- a socket set which covers the smaller range of nuts and bolts as well as 10 mm & 14 mm spark plug sockets
- a set of open ended spanners
- a range of screwdrivers
- a copper faced hammer or wooden mallet
- various sizes of drifts
- a quantity of paraffin, a 3/4" or 1" paint brush and some rags
- an aerosol of silver cellulose paint
- a tin of 'brush on black cellulose paint and a small brush

A rotor/stator tester can be purchased as a specific piece of testing equipment from a specialist supplier (usually the spare parts supplier stocks this item). However there are cheaper alternatives once it is realised that a rotor/stator tester is basically a circuit tester but which uses a higher voltage (i.e. 36 volts) than the normal 1.5 volt from the battery inside the tester. Bearing this in mind a tester can be constructed once a source of higher voltage is obtained. This voltage can be achieved in a number of ways.

A transformer which will give this amount of voltage or 3 x 12 volt car batteries linked together in series to give 36 volts.

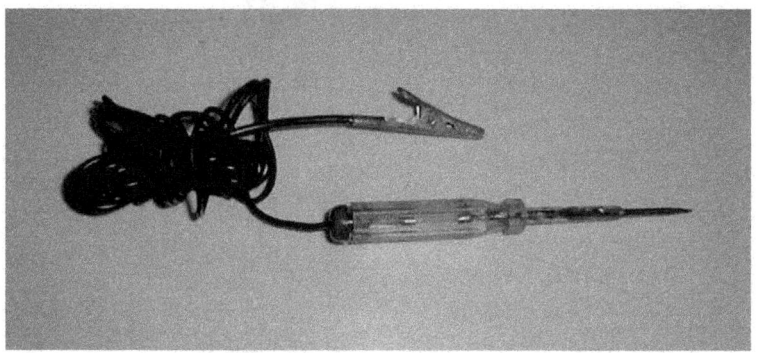

Building a rotor/stator tester:
Having secured the batteries together by means of self adhesive tape, the four 9 volt batteries need to be connected in series, that is each battery has its positive terminal connected to its neighbour's negative terminal by means of a soldered, insulated wire. Once completed this will leave the first battery's negative terminal and the last battery's positive terminal unattached. The multimeter now has to be incorporated into the arrangement in order to register a reading when a circuit is made. Set the multimeter to register direct current in the region of 36 volts without going off the dial register.

First by setting the multimeter to its highest DC range and putting the probes on the unattached end terminals of the battery pack; you should get a small reading. Reduce the DC range and repeat the test; the reading should be a little higher. Continue reducing the range and testing until the needle registers within the middle to top of the dial. The multimeter is then set. With the battery pack, solder a separate insulated wire to each of these two unattached end terminals, direct the one from the negative terminal to the multimeters negative input socket. Plug in one of the multimeter's probes into the positive input socket on the multimeter. This is now the negative probe. The insulated wire soldered to the positive end terminal of the battery pack is now the positive probe. To test the arrangement works, touch the probes together and a reading should register on the multimeter's dial.

Identifying the unit:
Each alternator is identifiable by a number stamped on it or on a label affixed to it. From this number it is possible to tell, through the relevant manufacturer's catalogue, which vehicles it will fit as well as identifying the correct spare parts required for its reconditioning.

STARTER MOTORS & ALTERNATORS

At first sight there seems to be a bewildering array of manufacturers of alternators on the market, Lucas, Delco Remey, Bosch, Paris-Rhone, Magneti Marelli etc. In order to simplify things it is suggested that one specific manufacturer be chosen as a base and the other manufacturers compared to it.

For instance, if you select Lucas as a base manufacturer, then all Lucas alternators would be identified within their main catalogue.

Once reconditioning has been completed it is recommended that a luggage label is tied to each unit showing its base manufacturer's reference number. This will minimise confusion later and act as a quick reference aid.

Main manufacturers' catalogues can be obtained through your local motor parts supplier or by directly contacting the manufacturing company concerned.

From an external point of view the differences between alternator models are as follows: - The width of the stator band i.e. 13 mm or 18 mm. Handing.

i.e. left hand mounting or right hand mounting. (The handing can be altered to suit during reconditioning.) The type and size of pulley fitted on the drive shaft including the number of grooves if it is a multi grooved type, the width of the belt if it is a single vee belt, the diameter of the pulley, as well as the overhang of the pulley from the main body. The type of connection terminal.

ALTERNATORS

The work of an alternator

When a car's engine is running it requires electrical energy to power the spark plugs which it gets from the battery, but if this were the only source of power the battery would soon run down as the charge was drained from it. To supplement this requirement for electrical power an alternator is mounted on the engine and, by means of a drive belt or fan belt, is rotated when the engine turns over. The alternator acts as a dynamo which, whilst it is turning, generates an electrical current. This is sent back to the battery to maintain its charge and ensure a constant flow of electrical power is available to power the spark plugs.

When the main shaft of the alternator is turned by the action of the drive belt from the engine, it causes the rotor, (which are the windings on the shaft), to rotate within the stator, (which are the fixed windings inside the casing). This arrangement acts as a dynamo, generating a constant flow of electrical energy with which to keep the battery fully charged. However, initially this energy is produced in the form of an alternating current and as the battery is a power source of direct current, it can only receive a charge of a similar nature.

This means that prior to being sent back to the battery, the energy being generated needs to be converted into a direct current. This is achieved by passing the generated current through the rectifier, which is a number of diodes linked together. These even out the oscillations within the energy being produced to form a direct current compatible with the battery. A certain amount of electrical energy is required within the rotor at all times in order to activate a magnetic field around and within the rotor. This magnetic field is an essential ingredient when generating power and initially is activated by the electrical supply from the battery via the ignition switch.

However, once the ignition key is released and the switch springs back one notch, this contact with the battery is broken and has to be supplemented by the alternator's own power. This power is gained by diverting some of the current generated by the alternator, towards a number of diodes within the rectifier dedicated to this task. Emerging as a direct current from these diodes, this power is then channelled through the regulator which, as its name suggests regulates the power to a constant flow no matter how fast the rotor is turning. From the regulator the current goes back to the rotor via the brushes, to do its job of activating the magnetic field.

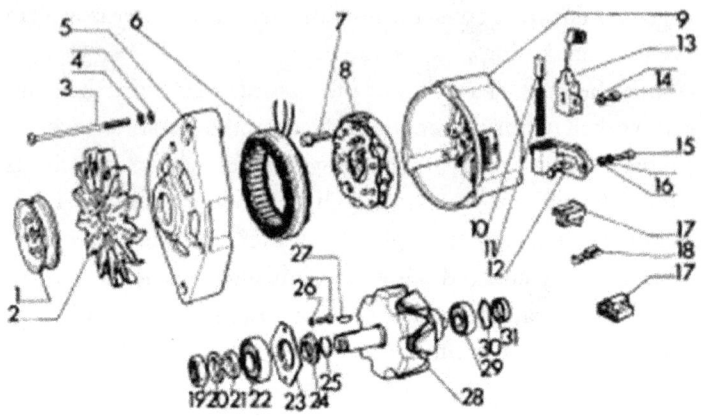

Main reasons for alternator malfunction:

The brushes have worn down. The brushes in an alternator are the means of electrical contact between a source of power, and the rotor's slip ring, in order to activate a magnetic field around and within the rotor, which is necessary to generate power. Accordingly, when these brushes wear down and contact becomes irregular or non-existent, then this magnetic field is interrupted, resulting in the disruption of power generation. The rectifier has failed The diodes within the rectifier are fundamental in converting alternating current into direct current. Any disruption in this process by means of a diode failure will affect the capacity of the alternator to perform its function, resulting in a reduction in the charge generated to supplement the drain on the battery. The regulator has failed The regulator has the job of stabilising the flow of current back to the rotor and any breakdown in the consistency of this flow will affect the activation of the magnetic field, the result is the same as brushes wearing down.

An initial stock is required from which to run the business and the best way to achieve this is to buy a quantity of used units from a scrap yard or vehicle dismantlers in order to recondition them. Try to ensure that all of the units are identifiable by means of a stamped reference number or attached label. Without it reference to its spare parts may be difficult although not always impossible. There may be a section in your main manufacturer's catalogue dealing with this problem particularly under casting numbers on drive end brackets.

Once you are familiar with the reconditioning procedure through renovating this initial stock, more specific purchases will have to be made to ensure an adequate conversion of enquires into orders is achieved. Because it is impossible to even contemplate stocking for every eventuality it is necessary to define a target in terms of vehicle manufacturer's types of models and years of manufacture. This target does not confine your ability to cope with other makes, models and years because as can be seen in the main manufacturers catalogues, most alternator models fit a considerable range of vehicles other than the chosen targets.

Alternator Parts

STARTER MOTORS & ALTERNATORS

Time will show which units need to be added to your stock and a note of enquiries that were not converted should be kept to assist in this.

Stripping down the alternator

Once the working area is set up with all the required equipment, work can begin in earnest stripping down an alternator. Although there are several different manufacturers of alternators on the market, the principles are basically the same and as long as the reasoning behind the procedure is understood then variations in parts can be dealt with as a matter of course.

Place the pulley in the engineer's vice and, with a spanner, loosen the nut attached to the shaft which is retaining the pulley and fan. Once loosened, take the whole assembly from the vice and continue to remove the nut whilst on the bench. If the nut is removed whilst in the vice it is possible that the main body could detach itself from the pulley and fall on the floor, possibly breaking a casting or even causing injury to yourself. Err on the side of caution and safety and finish disassembly on the bench. Place the nut and washer beneath it to one side.

The pulley can be detached from the shaft by sliding it off the end. If difficulty is encountered then a drop of penetrating oil and the use of a puller may assist in this operation. Place the pulley to one side.

Remove the woodruff key from the shaft, slide the fan off the shaft as well as the spacing collar and place them all to one side.

The alternator can now be placed on its end on top of the engineer's vice with the drive shaft being held firmly between the jaws. With the aid of an appropriate sized socket spanner. remove the bolts holding the plastic end cover in place. Place the cover and its retaining bolts to one side.

Make a note or sketch of all the wires on the now exposed end, noting their colours, where they emanate from and where they are connected to.

Remove the screws fitted to the brush housing top and place to one side.

The anti surge diode located alongside the rectifier which has a yellow lead, can be removed and placed to one side.

If a radio suppressor is fitted, remove it and place to one side.

Remove the screws retaining the regulator as well as the screw fixing its black earth lead to the end clamp. Place these to one side together with any connection strip associated with the regulator, noting its location.

Any further screws holding the brush housing in position will now be exposed and should be extracted before the housing itself is removed.

Take out the old brushes and with care, the one copper brush tension spring, noting which slide way it came from.

The rectifier is attached to the alternator in several ways and each one of these has to be disconnected. The three leads from the stator are soldered onto the outer plate of the rectifier. Apply a hot soldering iron to melt the solder on each one in turn and detach from the rectifier. There may be an earthing strip emanating from the inner rectifier plate and screwed to the end clamp. Remove the screw and place to one side. The whole assembly is bolted onto the end clamp. Loosen the outer nut and remove the rectifier placing it to one side.

Remove the bolts passing through one end clamp to the other and place to one side.

The main assembly now needs to be split by separating the end clamps and stator. If a slight twisting action has no effect then a wooden wedge gently tapped between the three pairs of tugs either side of the stator casing in alternate fashion should succeed. Place the stator and non drive end clamp to one side. The rotor now needs to be extracted from the drive end clamp. This is achieved by drawing the end clamp off the shaft with a puller.

Remove the spacer from the shaft of the rotor and place to one side. The rotor now needs to have the slip ring removed but before this can happen the rotor and slip ring need to be checked. With the circuit tester, place one probe on the copper inner ring and one probe on the outer ring and a reading should be seen. If no reading is obtained repeat the procedure but by placing the probes on the wires emanating from the solder spots on the side of the slip ring. If a reading is now observed this means the slip ring is faulty. If a reading is still not observed then the rotor is faulty and needs replacing.

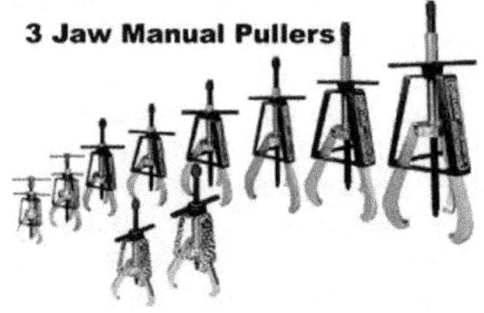

3 Jaw Manual Pullers

The rotor now needs to be extracted from the drive end clamp. This is achieved by drawing the end clamp off the shaft with a puller With the rotor/stator tester, place one probe on the copper inner ring of the slip ring and the other probe on the main shaft. There should be no reading indicating the insulation is OK. If there is a reading then the rotor is faulty and needs replacing. With a hot soldering iron melt the solder connections on the side of the slip ring and release the wires. Carefully slide off the slip ring and its spring clip. If the testing is OK and the faces of the slip ring are not excessively grooved or damaged, the slip ring can be put to one side for re-use. If there is any doubt as to the quality of the slip ring then it ought to be replaced. The bearing on the rotor now needs to be removed. The two wires which were soldered to the slip ring need to be pushed towards the shaft so that as the bearing is withdrawn they will not be damaged.

Alternator Bearings

With a bearing puller withdraw the bearing from the shaft taking care not to damage the plastic cover behind it. Mark the old bearing with an indelible pen to show it is used. It is easy to mix up old and new bearings and a little care to avoid a mistake at this point could prevent much greater problems later. Also remove the plastic dust cover, and as long as it is not damaged, place to one side. The rotor can also be placed to one side as long as it passed the tests.

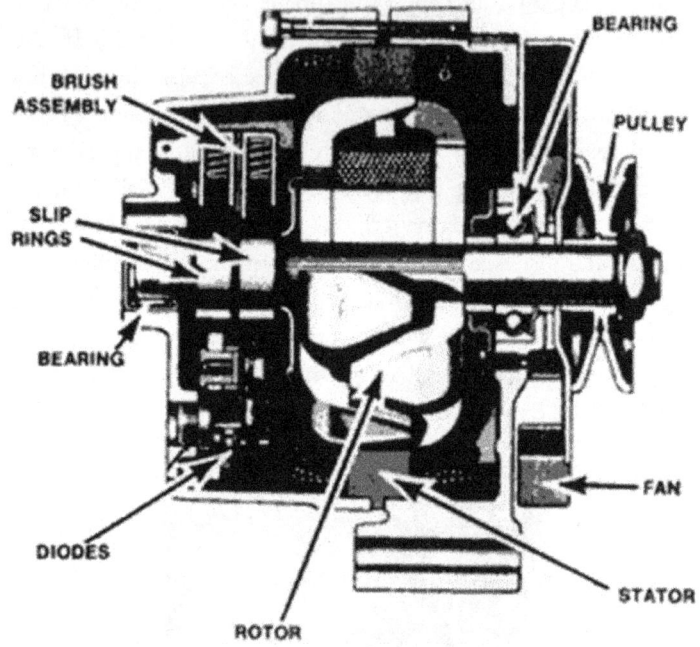

Disassembly is now complete.

The bearing assembly now needs to be removed from the inside face of the drive end clamp. This is done by first removing the circlip with a small screw driver; this will expose a metal flange which is removed and noted as the outer flange. Next the bearing itself should be removed followed by a rubber 0 ring and finally by an inner metal flange. With the exception of the bearing, which should be marked with an indelible pen and discarded, all the other items should be placed to one side with the drive end clamp.

Alternator Components

Obtaining spare parts

There are a number of companies who specialise in supplying spare parts for alternators and who are willing to do business with the smaller companies or individuals who are renovating or reconditioning these items. By far the best way to obtain parts is to search for suppliers on the internet or ask your local motor factors.

Testing and reconditioning alternators

Clean all of the components carefully with a paint brush and paraffin and allow to dry. Visually check the rotor and stator windings for damage or overheating. If any is visible then that particular component will have to be replaced. Place some newspapers on the work bench away from the disassembled components. Take the rotor and paint the two metal casings surrounding the windings with black brush on paint. Under no circumstances paint the windings or the shaft itself. Put to one side to dry.

Take the stator and, with the black brush on paint, paint the outer face only of the metal case (not the sides as these are required for testing at a later stage). Do not paint the windings, shaft or any other faces. Put to one side to dry.

One of the holes in the engine mounting lugs of one of the end clamps may be threaded. If it is, drill it out to the same diameter as the non threaded holes. This operation in effect gives the option of altering the handing of the alternator at a later stage without being detrimental.

Locate both end clamps, the fan and the pulley and spray with silver paint. When dry, turn over and spray the other side of each component.

A new bearing, together with the associated components, needs to be fitted into the drive end clamp next. The thinner of the two metal flanges is inserted first with its raised inner flange uppermost. Next fit the rubber O ring, the new bearing and then the other metal flange, again raised inner flange uppermost. Before the circlip can be fitted this whole assembly has to be compressed in order to give the circlip room to fit. Place the 14 mm spark plug socket onto the metal flange inside the end clamp. Put the circlip over the socket against the clamp ready to be fitted once compression takes place. Place this arrangement between the jaws of the engineer's vice and tighten. Fitting the circlip with the aid of a screwdriver is now quite easy as there is plenty of room for it to fit. Put this assembly to one side for now.

Place the rotor, drive shaft end down, on top of the vice with the jaws gripping the shaft itself. Slide on the plastic dust cover ensuring it is the right way round and does not foul the two wires running up the side of the shaft.

Ensuring that you have got the correct replacement bearing and again taking care not to foul the two wires running up the side of the shaft, place the new bearing on the shaft end. With the 10 mm spark plug socket, tap home the new bearing right up to the plastic dust cover. When this is located, bend the two wires to the sides ready for soldering.

The slip ring now requires fitting. This will either be the previous slip ring if its condition was considered good enough, or its replacement. Either way fit the spring clip to the slip ring first. Note that the end of the drive shaft has a groove across it and that inside the slip ring there are corresponding protrusions. Push on the slip ring and as it bottoms out twist it round to ensure those protrusions locate into the groove. The two wires bent to the side earlier should correspond to their location contacts on the slip ring.

Bend the two wires on the shaft up towards the slip ring where they should touch their contact points and solder the wires into place. Take the assembly out of the vice.

With the circuit tester place one probe on the copper inner ring and one probe on the outer ring and a reading should be seen. As long as it is successful you can proceed to the next step.

Assemble the rotor and drive end clamp together ensuring the spacer is located onto the shaft first. Take care not to damage the slip ring on the end of the shaft during this operation. Open the engineer's vice enough so that as you place the rotor onto the jaws, the slip ring and bearing are between the jaws out of harm's way. Using the 14 mm spark plug socket, you can now safely tap the drive end clamp onto the rotor shaft without causing any damage.

The stator now requires testing. With the circuit tester, check the three wires against each other by placing one probe on wire 1 and the other probe on wire 2; there should be a reading. Now check wire 1 against wire 3 and again there should be a reading. Finally check wire 2 against wire 3 and again there should be a reading.

You have now checked all the combinations possible to prove the wiring is sound. Should any of these readings fail then the stator will require replacing.

The stator can now be pushed onto the non drive end clamp ensuring that the three wires which are to be soldered onto the rectifier come through the appropriate gap in the end clamp.

Once the handing has been decided, line up the engine mounting bracket holes, push the drive end clamp onto the stator and fit the main assembly bolts. Check that the rotor turns freely within the assembly.

The alternator can now be placed on its end on top of the engineer's vice with the drive shaft being held firmly between the jaws.

A new rectifier now requires fitting. As this component is the main cause of failure in alternators, it is prudent to change it now even if it were to test OK, it has doubtless given long service and its life expectancy must be questionable. Ensuring the rubber plug is fitted to the back of the rectifier, slide it into place on the end clamp and tighten the nut. Refit any earthing strip that may have been fitted earlier.

Referring to the notes or sketches taken earlier before the alternator was stripped down, you can now solder in place the three wires emanating from the stator onto the rectifier. If in doubt refer to the old rectifier as you will be able to see where the previously used solder joints were located.

The brush housing can now be refitted. Replace the brush tension spring into its appropriate slide way and fit the new brushes. Screw down the brush retention strip screws without reconnecting any of the wires from the other components at this stage.

The regulator now needs fitting and again it is recommended that this component be renewed as it has spent some considerable time in service and will doubtless fail in due course if preventative action is not taken.

Refit the radio suppressor if one was fitted.

The anti surge diode can now be tested. With the circuit tester, place one probe against the metal fixing plate and the other probe against its wire; reverse the probes and repeat the test. There should be a reading in one direction only indicating that it's OK. If a reading in both or neither direction is made, then replace the diode. Fit the anti surge diode, or its replacement into its allotted position. It is now time to connect all of the components to their allotted contacts after first identifying the brush retention strips correctly. Carefully observe that one brush is positioned exactly in the centre of the end clamp and the other is offset to one side. These are known as the centre brush and outer brush respectively.

The regulator's connections
- fit its black wire to earth on the metal casing
- fit its yellow wire to outer brush retention strip
- fit its green wire to centre brush retention strip

If there is no green wire then a connection strip from the regulator fixing screw to the centre brush retention strip is required.

The anti surge diode's connection
- fit its yellow wire to the outer brush retention strip

The radio suppresser's connection
- fit its wire to the middle rectifier plate

The orange link's connections
- fit this wire from the outside rectifier plate to the outer brush retention strip

Should there be any other wires not covered by this list then refer to your notes taken before dismantling or ask your parts supplier for advice.

The fan and pulley can now be fitted to the end shaft. Refit the spacing collar and fan ensuring the fan is the correct way round. Next fit the washer and locking nut holding the pulley in a vice to ensure the nut is tight.

Variations in Alternator Models

The visual make up of the components within an alternator varies according to manufacturer. However, the principles are the same and once identification of each part is assured then the basic rules apply. In some makes of alternators the rectifier has only six diodes, leaving the other three as a separate component known as a trig diode or trio diode. The rectifiers and regulators vary considerably in appearance.

Rectifiers

Rectifiers - The basic rules to follow:

Regardless of make, always change the bearings, brushes, regulator and rectifier as well as the trig diode if one is fitted and change the slip ring if it proves faulty or well worn.

Bench testing the rebuilt alternator

It is assumed that at this stage you only have basic equipment and as such cannot perform the load test carried out by the professional reconditioner. However, having replaced both the regulator and the rectifier and performed all of the basic insulation and connection tests as well as having carefully followed the wire reconnection instructions, you can be reasonably certain that the alternator will function OK. It would be nice to be 100% certain that the alternator is in full working order but the expense of purchasing a professional motor driven bench mount and an alternator load tester is somewhat prohibitive. If you can afford, or feel the test equipment is necessary to allay your doubts, then the load tester and motor driven bench mount are available from your spare parts supplier together with full operating instructions. It is quite possible to construct your own motor driven bench mount for a fraction of the cost of the professional models. Although it may not perform all of the technical procedures of its big brother, (it is highly questionable that all these tests are necessary at the level you are working at anyway), it will give a very good indication as to whether the alternator is functioning correctly or not. An electric motor of 1 to 1.5 hp should be bolted to the bench. It should be fitted with a pulley which will complement the pulley on the alternators. A double lug fitting, suitable for bolting the alternators to, needs to be fitted to the bench in line with the motor at an appropriate distance away to allow a vee belt to straddle both the pulleys. The fact that the alternator can pivot on the double lug fitting means that tension can be applied to the belt. Once set up as described, the alternator needs to be wired up to a heavy duty battery in order to simulate the actual working arrangements in the car. When the electric motor is driving the alternator, the current reaching the battery should read 14 volts, indicating a charge is reaching the battery. Another more sophisticated test is to measure the amperage emanating from the alternator along the cable to the battery by means of a clamp meter. The amperage should correspond to that particular model's quoted output in the main manufacturer's catalogue.

Sale of Finished Goods

The first consideration to be made is, are you offering a fitting service or just selling exchange units? The answer to this question depends on the facilities you have at your disposal and your ability to undertake such tasks competently. Jacking up a car on your drive way or on the road in front of your house may well aggravate your neighbours or attract the attention of the local council. It may be that you have the use of a garage or work shop in which case fitting the units will not present a problem, but think carefully about this aspect before proceeding further. If it is your intention not to become involved in the fitting of these units then the public response will be limited and your sales efforts ought to be inclined towards the wholesale side of the business. If it is your intention to sell to the public then a simple marketing plan ought to be drawn up. A modest advert in the local paper on a regular basis should be considered. It has to be regularly inserted as Joe Public may have noticed the advert in the paper earlier but if, when he requires your services, your advert is not in that week's edition, he will have to go elsewhere. The yellow pages directory is a very good source of enquiries when assistance of this kind is required. However, orders for insertion are only accepted on an annual basis. Nowadays an online presence may well be more productive. Leafleting can also achieve good results but remember this work is seasonal to a certain extent. Although there is a steady demand for this service throughout the year, the busiest time is late autumn and early winter, just as the demands of colder, darker nights and heated rear screens take their toll after a summer of neglect. A leafleting campaign at this time will give the best returns. The local garages, auto electricians and repair centres (including the nationally advertised chains that are generally franchised) can also be a steady source of work as they have an established customer base constantly making enquiries. It is best to become acquainted with the owners or managers of these establishments yourself, as a personal assurance of your ability and sincerity will go a long way to persuade them to use you instead of their existing supplier.

  Finally, consideration ought to be given to the suppliers of parts to the motor trade, or Motor Factors as they are sometimes called, as in terms of turnover these outlets represent the biggest volume of trade by far. However, you will have to establish a trading track record as well as a good reputation based on quality as well as quantity before you could make inroads on any of these companies. In good time, and with the right effort, these things are quite possible if this is the direction you want to see yourself going. Above all, keep pushing your name in front of whoever will take notice. They may not want your products and services immediately but you can be assured at some point in time they will. Make your name synonymous with your product and ensure its you the customer thinks of first. It takes time to establish a reputation so don't be disheartened if it does not happen overnight, there will be a number of rejections before things take a positive turn for the better. The degree of success you achieve in the business will be a direct reflection of the effort you put in.

Pricing

On the basis that you are setting up in business as a starter motor and alternator repair and exchange service, you will likely need to purchase on a regular basis a number of reconditioned units for your customers which are not yet included in your stock. Approach a number of motor parts distributors for their comprehensive price lists, showing both trade and retail prices. You can then use this as a basis for your own pricing structure bearing in mind that your overheads should be low and you should be able to undercut the competition. Find out who your competitors are and ring them up for quotes as a member of the public. Get a number of prices for different models and set these against the prices from the parts distributors and you should be able to determine a pattern. For example, if the prices from your competitors are approximately 25% above the trade prices from the parts distributors or are a constant £10 above the trade prices then you will have determined their pricing pattern. You can then expect, with reasonable certainty, that extending the trade prices of the parts distributor by this factor will give you the rest of your competitors' prices, and as such can set your own to fall just below them. Do remember though when giving quotes to offer the units with and without fitting and to offer a warranty, usually 6 months. Prices to the garages are somewhat lower than those given to the public. In this instance you are competing directly with the motor parts distributors. You should have a trade price list. Initially set your prices between 10% and 15% lower than the distributors to give your potential customers a good incentive to buy from you. There should still be enough of a mark up on your reconditioned units cost prices to give a reasonable profit margin and anyway this differential can be addressed in the future once you become established.

Final considerations
One final consideration to be taken into account is a strong recommendation to take professional advice before becoming self employed, even on a part time basis. There are aspects that need to be considered at this point or they may become a financial burden at some future date. Public liability insurance in particular as well as tax, national insurance contributions and VAT are just a few of the issues that need to be addressed. A good accountant will usually give an individual a free first interview, without obligation, where he can discuss at length all of these things. He will also be able to give some indication of what his fees are likely to be should you decide to use his services. Remember the secret of success in business is to try and anticipate the inevitable problems that will arise and to plan how to deal with them before they happen.

STARTER MOTOR RECONDITIONING

History and Development of the Starter Motor
A hand crank was used to start engines, but it was inconvenient, difficult and dangerous to crank-start an engine. Even though cranks had an overrun mechanism, when the engine started the crank could begin to spin along with the crankshaft and potentially strike the person cranking the engine. Additionally, care had to be taken to retard the spark in order to prevent backfiring; with an advanced spark setting, the engine could kick back (run in reverse), pulling the crank with it, because the overrun safety mechanism works in one direction only.

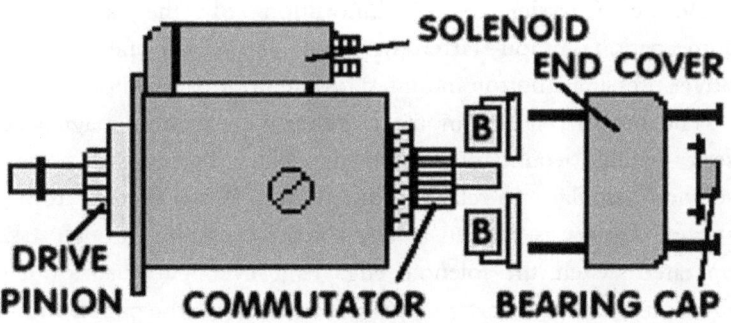

Although users were advised to cup their fingers under the crank and pull up, it felt natural for operators to grasp the handle with the fingers on one side, the thumb on the other. Even a simple backfire could result in a broken thumb; it was possible to end up with a broken wrist, or worse. Moreover, increasingly larger engines with higher compression ratios made hand cranking a more physically demanding endeavour.

While the need was fairly obvious, inventing one that worked successfully in most conditions did not occur until 1911 when Charles F. Kettering of Delco invented first useful electric starter. One aspect of the invention lay in the realisation that a relatively small motor, driven with higher voltage and current than would be feasible for continuous operation, could deliver enough power to crank the engine for starting. At the voltage and current levels required, such a motor would burn out in a few minutes of continuous operation, but not during the few seconds needed to start the engine. The starters were first installed by Cadillac on production models in 1912. These starters also worked as generators once the engine was running, a concept that is now being revived in hybrid vehicles. The Model T relied on hand cranks until 1919. By 1920 most manufacturers included self-starters, thus ensuring that anyone, regardless of strength or physical handicap, could easily start a car with an internal combustion engine.

Before Chrysler's 1949 innovation of the key-operated combination ignition-starter switch, the starter was operated by the driver pressing a button mounted on the floor or dashboard.

The modern starter motor is either a permanent-magnet or a series-parallel wound direct current electric motor with a starter solenoid (similar to a relay) mounted on it. When current from the starting battery is applied to the solenoid, usually through a key-operated switch, the solenoid engages a lever that pushes out the drive pinion on the starter driveshaft and meshes the pinion with the starter ring gear on the flywheel of the engine.

The solenoid also closes high-current contacts for the starter motor, which begins to turn. Once the engine starts, the key-operated switch is opened, a spring in the solenoid assembly pulls the pinion gear away from the ring gear, and the starter motor stops. The starter's pinion is clutched to its driveshaft through an overrunning sprag clutch which permits the pinion to transmit drive in only one direction. In this manner, drive is transmitted through the pinion to the flywheel ring gear, but if the pinion remains engaged (as for example because the operator fails to release the key as soon as the engine starts, or if there is a short and the solenoid remains engaged), the pinion will spin independently of its driveshaft. This prevents the engine driving the starter, for such back drive would cause the starter to spin so fast as to fly apart. However, this sprag clutch arrangement would preclude the use of the starter as a generator if employed in hybrid scheme mentioned above, unless modifications are made. Also, a standard starter motor is only designed for intermittent use which would preclude its use as a generator.

This overrunning-clutch pinion arrangement was phased into use beginning in the early 1960s, before that a Bendix drive was used. The Bendix system places the starter drive pinion on a helically-cut driveshaft. When the starter motor begins turning, the inertia of the drive pinion assembly causes it to ride forward on the helix and thus engage with the ring gear. When the engine starts, back drive from the ring gear causes the drive pinion to exceed the rotative speed of the starter, at which point the drive pinion is forced back down the helical shaft and thus out of mesh with the ring gear.

An intermediate development between the Bendix drive developed in the 1930s and the overrunning-clutch designs introduced in the 1960s was the Bendix Folo-Thru drive. The standard Bendix drive would disengage from the ring gear as soon as the engine fired, even if it did not continue to run. The Folo-Thru drive contains a latching mechanism and a set of flyweights in the body of the drive unit. When the starter motor begins turning and the drive unit is forced forward on the helical shaft by inertia, it is latched into the engaged position. Only once the drive unit is spun at a speed higher than that attained by the starter motor itself (i.e., it is back driven by the running engine) will the flyweights pull radially outward, releasing the latch and permitting the overdriven drive unit to be spun out of engagement. In this manner, unwanted starter disengagement is avoided before a successful engine start.

Chrysler Corporation contributed materially to the modern development of the starter motor. In 1962 Chrysler introduced a starter incorporating a gear train between the motor and the driveshaft. Rolls Royce had introduced a conceptually similar starter in 1946, but Chrysler's was the first volume-production unit. The motor shaft has integrally-cut gear teeth forming a drive gear which mesh with a larger adjacent driven gear to provide a gear reduction ratio of 3.75:1. This permits the use of a higher-speed, lower-current, lighter and more compact motor assembly while increasing cranking torque. Variants of this starter design were used on most vehicles produced by Chrysler Corporation from 1962 through to 1987. The Chrysler starter made a unique, readily identifiable sound when cranking the engine.

This starter formed the design basis for the offset gear reduction starters now employed by about half the vehicles on the road, and the conceptual basis for virtually all of them. Many Japanese automakers phased in gear reduction starters in the 1970s and 1980s. Light aircraft engines also made extensive use of this kind of starter because its light weight.

Those starters not employing offset gear trains like the Chrysler unit generally employ planetary epicyclic gear trains instead. Direct-drive starters are almost entirely obsolete owing to their larger size, heavier weight and higher current requirements. Ford also issued a nonstandard starter, a direct-drive movable pole shoe design that provided cost reduction, rather than electrical or mechanical benefits. This type of starter eliminated the solenoid, replacing it with a movable pole shoe and a separate starter relay.

The Ford starter operated as follows:
- The operator closed the key-operated starting switch.
- A small electric current flowed through the starter relay coil, closing the contacts and sending a large current to the starter motor assembly.
- One of the pole shoes, hinged at the front, linked to the starter drive, and spring-loaded away from its normal operating position, swung into position. This moved a pinion gear to engage the flywheel ring gear, and simultaneously closed a pair of heavy-duty contacts supplying current to the starter motor winding.
- The starter motor cranked the engine until it started. An overrunning clutch in the pinion gear uncoupled the gear from the ring gear.
- The operator released the key-operated starting switch, cutting power to the starter motor assembly.
- A spring retracted the pole shoe, and with it, the pinion gear.

This starter was used on Ford vehicles from 1973 onwards, then a gear-reduction unit conceptually similar to the Chrysler unit replaced it. Light motor vehicles have now adopted 9.6 volt to 10.4 volt starter motors for use with 12 volt systems to give increased power. The lower current starter will give increased torque, but will tend to overheat and burn out with prolonged use under load.

Some diesel engines from 6 to 16 cylinders are started by means of a hydraulic motor. Hydraulic starters and the associated systems provide a sparkless, reliable method of engine starting at a wide temperature range. Typically hydraulic starters are found in applications such as remote generators, lifeboat propulsion engines, offshore fire pumping engines, and hydraulic fracturing rigs. The system used to support the hydraulic starter includes valves, pumps, filters, a reservoir, and piston accumulators. The operator can manually recharge the hydraulic system; this cannot readily be done with air or electric starting systems, so hydraulic starting systems are favoured in applications wherein emergency starting is a requirement.

The repair and reconditioning of starter motors can be a highly lucrative business. We will discuss how you can get started in this business, what equipment you need and how to develop the necessary skills.

SERVICE PARTS

1 - SOLENOID
2 - BRUSH BOX ASSEMBLY
3 - BRUSH KIT
4 - ARMATURE
5 - RETAINTION KIT
6 - PIVOT AND GROMET KIT
7 - DRY SHAFT AND BEARING BRACKET ASSEMBLY KIT
8 - DRIVE ASSEMBLY
9 - BUSH KIT
10 - MOTOR ASSEMBLY
    DRIVE SHAFT
    SUNDRY KIT
    SUNDRY PARTS KIT

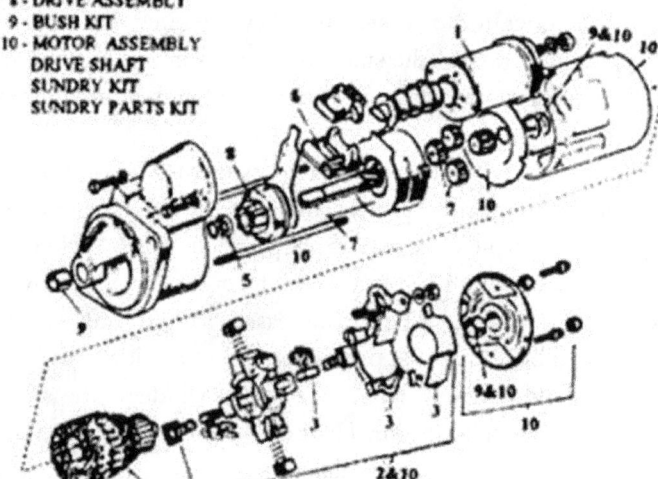

Equipment needed for reconditioning starter motors:

- A good working area. A garage, shed or even a spare room will suffice but one that offers adequate space to store finished products, preferably on some type of shelving system for ease of access.
- A work bench at a height suitable for working in a seated position.
- A pair of circlip pliers.
- A main manufacturer's starter motor catalogue.
- A main manufacturer's cross reference catalogue.
- Spare parts manual.
- A variable range circuit tester or multi meter.
- A pair of good quality wire cutters.
- An engineer's vice.
- An adjustable bench lamp.
- A soldering iron, solder and flux.
- Socket set which covers the smaller range of nuts and bolts.
- A pair of mole grips.
- A set of open ended spanners.
- A range of screwdrivers.
- A 12 volt battery.
- Set of jump leads.
- A short length of heavy duty insulated cable with an eyelet on one end and a female spade fitting on the other.
- A copper faced hammer or wooden mallet.
- Various sizes of drifts.
- A quantity of paraffin, a 3/4' or 1" paint brush and some rags.
- Two aerosols of cellulose paint, one of black and one of silver.
- Masking tape.

STARTER MOTORS & ALTERNATORS

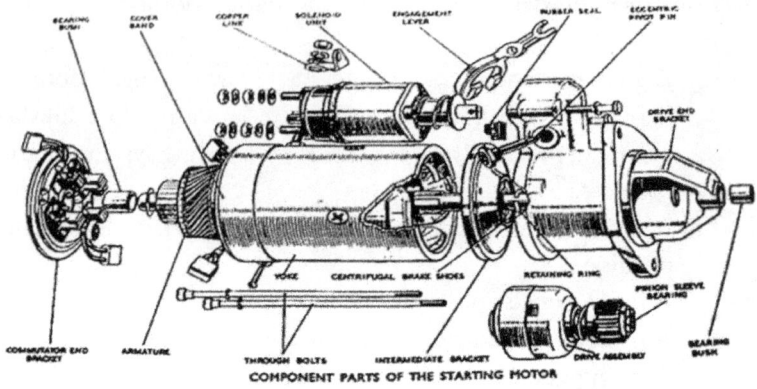

COMPONENT PARTS OF THE STARTING MOTOR

Identifying the unit

Each starter motor is identifiable by a number stamped on it or on a label affixed to it. From this number it is possible to tell, through the relevant manufacturer's catalogue, which vehicles it will fit as well as identifying the correct spare parts required for its reconditioning.

There are many different manufacturers of starter motors on the market. Lucas, Delco Remey, Bosch, Paris-Rhone, Magneti Marelli etc. To simplify things choose one specific manufacturer as a base and the other manufacturers compared to it. If for instance Lucas is our base manufacturer, then all Lucas starter motors would be identified within their main catalogue.

The catalogue tackles identification in a number of ways. Firstly it lists all past and present vehicle makes and models and shows which Lucas unit fits each particular model. Next, each Lucas unit is listed numerically and shows which vehicles and models that particular unit is capable of fitting. Finally, a comprehensive list of competitor's units identification numbers is given with the Lucas equivalent along side. This means that although you may have a unit manufactured by a company other than Lucas, from this catalogue you can still tell which vehicles it is capable of fitting.

Once reconditioning has been completed tie a luggage label to each unit showing its base manufacturer's reference number. This will minimise confusion later and act as a quick reference aid.

Main manufacturers' catalogues may be obtained through your local motor parts supplier or by directly contacting the manufacturing company concerned.

The work of a starter motor
A starter motor is a powerful electric motor which, when briefly connected to the battery, (which is what happens when the ignition key is turned and held in the start position) engages onto the engine by means of gearing and turns the engine over allowing it to fire into life. Once the engine fires, the ignition key is released allowing it to spring back one notch, thus disengaging the battery from the starter motor and allowing it to come to rest.

The starter motor needs to be very powerful as turning an engine over requires a lot of energy. The drain on the battery is very great indeed. Again, anyone who has tried to start a car which for some reason will not fire, knows only too well that the battery soon runs flat.

As a result there is a balance of requirements needed from the manufacturers. Not enough power in the starter motor will result in the engine not being turned over or turned over too slowly. Too much power in the starter motor will result in excessive power drainage of the battery.

To answer this problem car manufacturers calculate the energy required to rotate the engine of each of their vehicle models. A range of starter motors of varying power, suitable to their particular make may then be selected in order to match the energy requirements of each model,

Basically, when an electrical current is applied to a starter motor a centre shaft, known as an armature, is made to rotate within its round metal casing. Attached to one end of this shaft is a small gear cog which, when engaged to the engine by means of the ring gear, turns the engine over in order to make it start.

However, when not in use this gear cog needs to be free of the engine. This is achieved by means of the solenoid which is the small cylindrical component attached to the outside of the starter motor. The job of this component is to move the gear cog sideways along the armature thus engaging the ring gear just before the current is applied to the starter, and to return it to its neutral position when the current is disconnected.

Sequence of events when a starter motor is activated:
The solenoid is a powerful electromagnet surrounding a plunger. The plunger has a toggle mechanism called a pivot arm attached to one end which in turn is attached to the sliding gear cog on the armature. When current is applied to the starter motor the electromagnet in the solenoid pushes the plunger deep inside itself to a stopped end, the pivot arm attached to the end of the plunger is pulled in and hinges around a fixed pivot, pulling the gear cog on the armature out along its length to engage with the ring gear on the engine.

One further feature of the solenoid is that as the electrical current is applied, the plunger has to travel down inside the solenoid before it makes a further electrical contact at the far end allowing the current to continue on to activate the armature. This ensures that the gearing is engaged before the starter motor begins to turn over.

When the current finally reaches and activates the armature to act as an electric motor, it turns very fast, forcing the engine to turn over and fire into life. At this point the small gear on the armature has a ratchet mechanism built into it very similar to a free wheel on a bicycle. Should the engine fire into life and rotate the gearing at a faster pace than the starter motor wants to go, then the gear will turn on the armature faster than the armature (free wheeling) without causing any damage to the starter motor.

When the current is disengaged the armature stops turning, the electromagnet ceases to hold the plunger in and at the same time a compressed spring within the solenoid returns the plunger to its out position. This in turn activates the toggle around its fixed pivot in the opposite direction thus disengaging the small gear cog from the engine ring gear.

Main reasons for starter motor malfunction:

- The brushes have worn down

The brushes in a starter motor are the means of electrical contact between the cable from the battery, and the rotating armature inside the motor. They are pushed against the armature by means of small springs in order to maintain contact. As a result friction eventually wears down these brushes to such an extent that the springs can no longer force contact to be made with the armature and a failure occurs.

The solenoid has developed a fault
When the electromagnet has drawn the plunger into the solenoid to activate the pivot arm, its second job is to act as a switch, allowing a large amount of current to travel through its contacts and on to the armature to turn the motor. As such, just before contact is made and again just after contact is broken, arcing occurs, which is a spark jumping across the small gap between the two contacts. This is quite normal but it does have the effect of creating a small burn or carbonisation on each occasion. Over a period of time the effectiveness of the contacts to touch each other and pass the current along dimin-ishes.

The resulting sound of a single click coming from the engine compartment when the ignition key is turned is synonymous with this fault. It signifies that the plunger in the solenoid has been activated but when it has reached the end of the cylinder the contacts to transfer the current to the armature are burnt and the current does not reach the armature.

These two faults represent the majority of malfunctions in starter motors and are quite easy to deal with.

Building up stocks

An initial stock is required from which to run the business and the best way to achieve this is to buy a quantity of used units from a scrap yard or vehicle dismantlers in order to recondition them.

Once you are familiar with the reconditioning procedure, through renovating this initial stock, more specific purchases will have to be made to ensure an adequate conversion of enquires into orders is achieved. Because it is impossible to even contemplate stocking for every eventuality it is necessary to define a target in terms of vehicle manufacturers, types of models and years of manufacture. This target does not confine your ability to cope with other makes, models and years because, as can be seen in the main manufacturers catalogue, most starter motors models fit a considerable range of vehicles other than the chosen targets.

Stripping down the starter motor

Once the working area is set up work can begin in earnest stripping down a starter motor. Although there are a number of different manufacturers of starter motors on the market, the principles are basically the same and as long as the reasoning behind the procedure is understood, then variations in parts can be dealt with as a matter of course.

Procedure

Make a note or sketch of the solenoid end cap and in particular which connection is fitted to the main body. Disconnect this lead Disconnect the solenoid from the drive end bracket by means of the nuts or screws holding it in place.

The main body of the solenoid can now be pulled away leaving the plunger still attached to the pivot arm. Care should be taken not to lose the plunger return spring which is now loose within or around the plunger. Place the solenoid to one side.

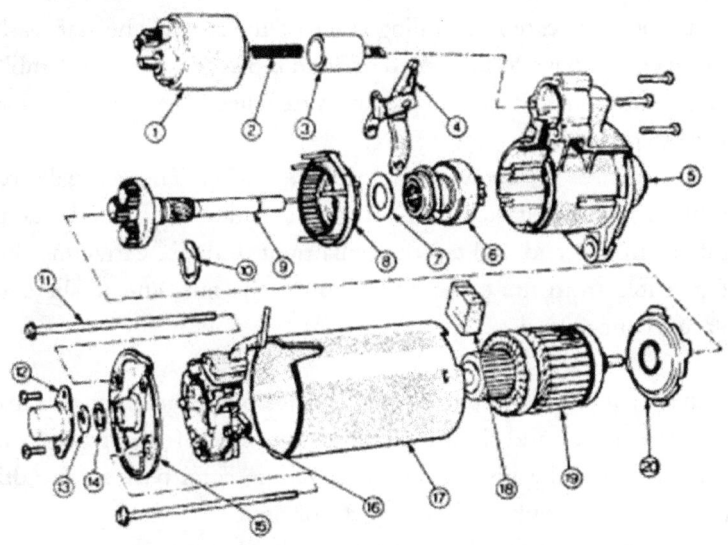

Starter Motor Components

At this point it has to be determined what type of pivot mechanism is employed on the pivot arm. If a pivot pin can be seen passing through either side of the drive end bracket then it needs to be removed. If it is held in place by a small split pin or circlip then remove it and the pivot pin will tap out and should be put to one side. If there is no split pin or circlip then the pivot pin is usually expanded at one end and can be carefully drilled out. If no pivot pin is visible then disregard this step as the pivot arm will be internally pivoted within the end bracket.

Remove the dust cover from the non drive end of the main body. This may be a push fit or attached by screws. Place to one side.

The removal of the dust cover will reveal the end of the armature. This is usually held in position by means of a star lock washer or circlip. Whichever system has been used it needs to be removed. The circlip will easily pull out to one side with the aid of a screw driver, but take care not to fling it across the room. The star washer however, will have to be forced off and replaced when reassembled. Also remove any shims or washers, note where they came from and place to one side.

The main body screws now need removing. These usually pass right through the starter motor from one end to the other. Undo the nuts from one end and tap the bolts right through, extracting them if possible from the other end. Replace the nuts and washers and place to one side.

If the pivot pin was removed, then at this point the pivot arm itself will also be free and should be put to one side. If no pivot pin was removed then the pivot arm will be an integral part of the drive assembly and should be left as such for now.

The non drive end bracket may be held in position by its own screws. If it is remove them, if it is not then the main body screws would have been holding it in place. Either way, a gentle tap will now loosen it from the main body. Do not fully separate yet.

At this point observe the inside of the non drive end bracket and make a note or sketch showing which brushes go into which slide ways and make a reference to those slide way positions in relation to the main body screws.

Carefully extract the two brushes that are attached to the field coils. This will release the non drive end bracket from the main body. Place the main body to one side.

Remove the brass bushes from both the drive end and non drive end. This can be done by means of gently tapping an appropriate sized drift against the end of the bush. This drift needs to be small enough to pass easily through the bush housing aperture but big enough to push the bush out. Mark these old bushes with an indelible marker to show they are used. It is easy to mix up old and new bushes and a little care to avoid a mistake at this point could prevent much greater problems later.

Disassembly is now complete.

Obtaining spare parts
There are a number of companies in the market who specialise in supplying spare parts for starter motors and who are willing to do business with the smaller companies or individuals who are renovating or reconditioning these items. It is worth noting that from an initial enquiry being registered with the supplier, it may take some time before their representative makes contact.

STARTER MOTOR TESTING

With the exception of the drive assembly, clean all of the components carefully with a paint brush and paraffin and allow to dry. The drive assembly has to be treated separately as it has sealed for life lubrication within its workings, which can be affected by the liberal use of paraffin. The best way to treat this component is to lightly moisten it with some paraffin on a rag and to carefully wipe away any dust, dirt or grease from the surface. Then wipe dry with a clean rag.

With the circuit tester, check the insulation between the brush springs and the non drive end bracket. Place one probe on the unpainted end bracket metal and the other probe on one of the springs. There should be no reading, indicating that the insulation is good.

Test each spring in turn. If a reading occurs on any of the springs then the insulation is faulty and will need replacing. As the brush housings are usually riveted onto the end bracket the rivets will have to be drilled out and the insulation material renewed (gasket paper is an adequate material). Re-rivet the brush housing back into place.

If there is not, then the field coil is faulty and will have to be replaced, but this is a very rare occurrence. Replacement of the field coil is achieved by removing the four pole screws from the outer casing to free the pole shoes and thus releasing the field coil.

Whether the brushes are worn down or not it is good practice to always replace these components. Start with the brushes attached to the end bracket and ensure that the new brushes are the correct replacements. Take one of the old brushes out of its slide way and cut the copper cable about half way along its length with a pair of wire cutters. Ensure the new brush has a similar length of copper cable to that on the old brush (trim to length if necessary) and carefully solder the ends of the copper cable together. Repeat the procedure for the other brush. Put the new brushes into their appropriate slide ways, making sure the copper cable is coming out of that part of the brush closest to the spring end. With both brushes renewed, and positioned in their slide ways, check visually that the copper cable does not or will not come into contact with the end bracket casing. If in doubt apply insulation tape around the cable. The wiring within the field coil has to be checked to ensure that it is sound and there are no breakages. With the circuit tester, place one probe on the lead which was connected to the solenoid, and the other probe on one of the brushes attached to the field coil. There should be a positive reading indicating current flow.

The two brushes attached to the field coil have to be replaced. This time bend the metal strip on the inside of the housing down to reveal where the copper cable has been welded into place. At this point make a note which is the longer cable and which is the shorter and cut both cables about 5 mm from the metal strip. Slide the insulation tube off the old brush cables and, if it is in good order, use it on the new brushes. If in doubt, replace the tube. Cut the copper cables on the new brushes to match the length of the old ones and slide on the insulation tube. Having previously taken note which side of the metal strip the long and short cables go, fit the new brushes accordingly by soldering them in place. With the first brush hold the insulation tube back along the cable with a pair of mole grips. The new brush cable end can now be soldered to the stub of the old cable left on the metal strip. Repeat the procedure with the shaft for wear or damage and replace if necessary.

If you do have to take the pinion and drive mechanism off the armature for any reason then the following procedure should be followed.

- Place the armature, commutator end down in a previously drilled hole in the work bench, having first placed some protective cloth beneath it.
- Slide the mechanism down towards the windings as far as it will go.
- The now exposed ring stop should be tapped down along the shaft.
- A spring ring held in a groove on the shaft will now be exposed and this should be taken off the shaft.
- The whole pinion and drive mechanism can now be removed.
- From the mechanism, remove the circlip and washer retaining the pivot arm or its location lugs.
- Remove the pivot arm or its location lugs along with any associated parts.
- The pivot arm needs checking for damage or wear, as does the pivot pin if one was fitted. If in doubt then replace these parts.

The solenoid now requires reconditioning by replacing the end cap. Remove the end cap screws and, with a soldering iron, melt the solder contact joints. There are usually two but there may be three, depending on the model. Once the joints have been melted, and the old solder removed, the end cap can be taken off and discarded. The moving copper contact, which has now been exposed, should be taken out and if there is no replacement clean its contact spots with an emery cloth to remove any pitting or carbonisation. Having replaced the moving copper contact with either a new one or the now cleaned old one, fit the new end cap in position ensuring that the appropriate wires are coming through their allotted holes. Solder these wires into position and replace the end cap screws, terminal nuts and washers. If it is not possible to renew just the end cap then it is recommended that the whole solenoid be replaced.

Ensuring that all of the commutator end shims or washers have been replaced on the armature shaft, this assembly can now be replaced within the main body housing and the non drive end bracket which was painted earlier. As the shaft end passes through the end plate replace any previously fitted washers and circlip or fit a new star washer.

All of the brushes have to be pushed back against their springs to allow the drum to pass into the brush box before the end bracket can be put in place. Again, replace any washers and circlip or fit a new star washer. Refit the dust cover.If the pivot arm was located in the drive end bracket by means of a pivot pin then this should be refitted at this stage. Position the pivot arm in the drive end bracket so that the pivot pin holes are aligned. Slide the pivot pin through the holes and secure it as required. This may be with a split pin or circlip, if the pin has a hollow end, spread with a rivet punch. Fit the drive end bracket onto the assembly ensuring the pivot arm is located properly on to the pinion drive unit. The main housing screws or bolts can now be inserted and tightened.

Locate the solenoid plunger onto the pivot arm and test that the pivot arm moves freely back and forth, moving the drive mechanism along its helicoiled shaft. Ensure the plunger return spring is in position.

Referring to the notes or sketch made earlier when the solenoid was removed, ensure the solenoid main body is the correct way round. Slide the plunger into the main body and refit it to the drive end bracket by means of the nuts or screws previously removed.

Refit the lead from the main body to the correct terminal on the solenoid.

Bench testing the rebuilt starter motor

The 12 volt battery needs to be positioned near to the engineer's vice. Place the starter motor in the vice so that the metal faces of the vice come into contact with the main body. Fit the short insulated cable to the spade, or smaller terminal on the solenoid end cap. Leave its other end well away from the starter or vice ensuring it cannot come into accidental contact. With the red jump lead, connect the positive terminal of the battery to the outer most large terminal on the solenoid end cap.

With the black jump lead connected to the negative terminal of the battery, touch the other end against the metal vice; there should be no reaction. If the motor starts to turn then there is a fault within the solenoid end cap which has probably been replaced incorrectly. As long as there is no reaction, clamp the black negative lead to the vice.

Take the loose end of the lead which is connected to the spade, or smaller terminal of the solenoid end cap and touch it against the terminal holding the red lead from the positive terminal of the battery. The starter should come to life, first moving the drive pinion along the shaft and then rotating the armature. Finally, when the lead is disconnected, the drive pinion should retract along the shaft and the armature come to rest.

Fit a pair of mole grips to the drive end gear cog so that it is held securely and cannot rotate. Repeat the last procedure to check the quality of the ratchet mechanism within the drive pinion. As long as the armature does not rotate then the ratchet mechanism is Ok.

Having reached this point successfully the black paint can be touched up where it was held in the vice and a luggage label attached indicating its base manufacturer's reference number. The starter motor has now been successfully reconditioned and can be put on the shelf ready for sale

SALE OF FINISHED GOODS

The first consideration to be made is are you offering a fitting service or just selling exchange units? The answer depends on the facilities you have at your disposal and your ability to undertake such tasks competently. It may be that you have the use of a garage or work shop, in which case fitting the units will not present a problem but think carefully about this aspect before proceeding further. If it is your intention not to become involved in the fitting of these units then the public response will be limited and your sales efforts ought to be inclined towards the wholesale side of the business.

If it is your intention to sell to the public then a simple marketing plan ought to be drawn up. A modest advert in the local paper on a regular basis should be considered. It has to be regularly inserted as customers may have noticed the advert earlier but if, when he requires your services, your advert is not in that week's edition, he will have to go elsewhere.

Of course an online website is essential and if you have the necessary simple skills this can be made at little or sometimes no cost.

The yellow pages directory is a very good source of enquiries as this is usually the first place anyone looks when assistance of this kind is required. However, orders for insertion are only taken on an annual basis.

Leafleting the local garages, auto electricians and repair centres (including the nationally advertised chains that are generally franchised) can also be a steady source of work as they have an established customer base constantly making enquiries. It is best to become acquainted with the owners or managers of these establishments yourself as a personal assurance of your ability and sincerity will go a long way to persuade them to use you instead of their existing supplier.

Finally, consideration ought to be given to the suppliers of parts to the motor trade, or motor factors as they are sometimes called, as in terms of turnover these outlets represent the biggest volume of trade by far. However, you will have to establish a trading track record as well as a good reputation based on quality as well as quantity before you could make inroads on any of these companies. In good time and with the right effort these things are quite possible if this is the direction you want to see yourself going.

Above all, keep pushing your name in front of whoever will take notice. They may not want your products and services immediately but you can be assured at some point in time they will. Make your name synonymous with your product and ensure it's you the customer thinks of first.

It takes time to establish a reputation so don't be disheartened if it does not happen overnight, there will be a number of rejections before things take a positive turn for the better. The degree of success you achieve in the business will be a direct reflection of the effort you put in.

Pricing
On the basis that you are setting up in business as a starter motor and alternator repair and exchange service, and as such you are likely to want to purchase on a regular basis, a number of reconditioned units for your customers which are not yet included in your

stock. Approach a number of motor parts distributors for their comprehensive price lists showing both trade and retail prices. You can then use this as a basis for your own pricing structure bearing in mind that your overheads should be low and as such, you should be able to undercut the competition.

Find out who your competitors are and ring them up for quotes as a member of the public. Get a number of prices for different models and set these against the prices from the parts distributors and you should be able to determine a pattern. For example, if the prices from your competitors are approximately 25% above the trade prices from the parts distributors or are a constant £10 above the trade prices then you will have determined their pricing pattern. You can then expect, with reasonable certainty, that extending the trade prices of the parts distributor by this factor will give you the rest of your competitors' prices, and as such can set your own to fall just below them. Do remember though when giving quotes to offer the units with and without fitting and to offer a warranty, usually six months.

Prices to the garages are somewhat lower than those given to the public. In this instance you are competing directly with the motor parts distributors and as such, if you followed the advice given earlier in this section, should have a trade price list. Initially set your prices between 10% and 15% lower than the distributors to give your potential customers a good incentive to buy from you. There should still be enough of a mark up on your reconditioned units cost prices to give a reasonable profit margin and anyway this differential can be addressed in the future once you become established.

Final considerations
One final consideration to be taken into account is a strong recommendation to take professional advice before becoming self employed, even on a part time basis. There are aspects that need to be considered at this point or they may become a financial burden

at some future date. Public liability insurance in particular as well as tax, national insurance contributions etc are just a few of the issues that need to be addressed.

A good accountant will usually give an individual a free first interview, without obligation, where he can discuss at length all of these things. He will also be able to give some indication of what his fees are likely to be should you decide to use his services.

Remember the secret of success in business is to try and anticipate the inevitable problems that will arise and to plan how to deal with them before they happen.

VEHICLE BATTERIES

No manual of repairing starter motors and alternators would be complete without a short description of vehicle batteries. After all, many alternator and starter problems can be due to a faulty battery, and vice versa.

An automotive battery is a type of rechargeable battery that supplies electric energy to an automobile. Usually this refers to an SLI battery (starting, lighting, ignition) to power the starter motor, the lights, and the ignition system of a vehicle's engine. This also may describe a traction battery used for the main power source of an electric vehicle.

Automotive batteries (usually of lead-acid type) provide a nominal 12-volt potential difference by connecting six galvanic cells in series. Each cell provides 2.1 volts for a total of 12.6 volt at full charge. Lead-acid batteries are made up of plates of lead and separate plates of lead dioxide, which are submerged into an electrolyte solution of about 35% sulphuric acid and 65% water. This causes a chemical reaction that releases electrons, allowing them to flow through conductors to produce electricity. As the battery discharges, the acid of the electrolyte reacts with the materials of the plates, changing their surface to lead sulphate. When the battery is recharged, the chemical reaction is reversed: the lead sulphate reforms into lead oxide and lead. With the plates restored to their original condition, the process may now be repeated.

Lead-acid batteries for automotive use are made with slightly different construction techniques, depending on the application of the battery. The typical battery in use today is of the flooded cell type, indicating liquid electrolyte. AGM or absorbed glass mat type batteries have electrolyte immobilised as a gel.

This deals with the flooded type of car battery. The starting (cranking) or shallow cycle type is designed to deliver large bursts of energy, usually to start an engine. The SLI batteries usually have a greater plate count in order to have a larger surface area that provides high electric current for short period of time. Once the engine is started, they are recharged by the engine-driven charging system.

The deep cycle (or motive) type is designed to continuously provide power for long periods of time (for example in a trolling motor for a small boat, auxiliary power for a recreational vehicle, or traction power for a golf cart or other battery electric vehicle). They can also be used to store energy from a photovoltaic array or a small wind turbine. They usually have thicker plates in order to have a greater capacity and survive a higher number of charge/discharge cycles. The specific energy is in the range of 30-40 watt-hours per kilogram.

Batteries intended for starting, lighting and ignition (SLI) systems are intended to deliver a heavy current for a short time, and to have a relatively low degree of discharge on each use. They have many thin plates, thin separators between the plates, and may have a higher specific gravity electrolyte to reduce internal resistance. Deep-cycle batteries have fewer, thicker plates and are intended to have a greater depth of discharge on each cycle, but will not provide as high a current on heavy loads.

Some battery manufacturers claim their batteries are dual purpose (starting and deep cycling). Some cars use more exotic starter batteries, the 2010 Porsche 911 GT3 RS offers a lithium-ion battery as an option to save weight over a conventional lead-acid battery.

Car batteries using lead-antimony plates would require regular watering top-up to replace water lost due to electrolysis on each charging cycle. By changing the alloying element to calcium, more recent designs have lower water loss unless overcharged. Modern car batteries have reduced maintenance requirements, and may not provide caps for addition of water to the cells. Such batteries include extra electrolyte above the plates to allow for losses during the battery life. If the battery has easily detachable caps then a top-up with distilled water may be required from time to time. Prolonged overcharging or charging at excessively high voltage causes some of the water in the electrolyte to be broken up into hydrogen and oxygen gases, which escape from the cells. If the electrolyte liquid level drops too low, the plates are exposed to air, lose capacity, and are damaged. The sulphuric acid in the battery normally does not require replacement since it is not consumed even on overcharging. Impurities or additives in the water will reduce the life and performance of the battery.

Manufacturers usually recommend use of demineralised or distilled water since even potable tap water can contain high levels of minerals.

Charge and discharge
In normal automotive service the vehicle's charging system powers the vehicle's electrical systems and restores charge used from the battery during engine cranking. When installing a new battery or recharging a battery that has been accidentally discharged completely, one of several different methods can be used to charge it. The most gentle of these is called trickle charging. Other methods include slow-charging and quick-charging, the latter being the harshest. In most cars, the voltage regulator of the charge system is unaware of the relative currents charging the battery and for powering the car's

loads such as engine control, fans and lighting.

The charge system essentially provides a fixed voltage of typically 13.8 to 14.4 V (Volt), unless the alternator is at its current limit. A discharged battery draws a high current of typically 20 to 40 A (Ampere).

As the battery gets charged the charge current typically decreases 2 A to 5 A. A high load results when multiple high-power systems such as radiator fan, heater blowers, lights and entertainment system are running. In this case, the battery voltage will decrease and the charge current as well. Some manufacturers include a built-in hydrometer to show the state of charge of the battery. This acrylic eye has a float immersed in the electrolyte. When the battery is charged, the specific gravity of the electrolyte increases (since all the sulphate ions are in the electrolyte, not combined with the plates), and the coloured top of the float is visible in the window. When the battery is discharged (or if the electrolyte level is too low), the float sinks and the window appears yellow (or black).

The built-in hydrometer only checks the state of charge of one cell and will not show faults in the other cells. In a non-sealed battery each of the cells can be checked with a portable or hand-held hydrometer.

Batteries will last longer if not stored in a discharged state.

In emergencies a vehicle can be jump started, by the battery of another vehicle or by a hand portable battery booster. Whenever the car's charge system is inadequate or too slow to fully charge the battery, a battery charger can be used. Simple chargers will not regulate the charge current and the user needs to stop the process or lower the charge current to prevent excessive gassing of the battery.

More elaborate chargers charge the battery fully and safely in a short time without requiring user intervention. Storage batteries should be monitored and periodically charged if in storage to retain their capacity. Batteries intended to be stored should be fully charged, cleaned of corrosion deposits, and left in a cool dry environment. High temperatures increase the self discharge rate and plate corrosion. Lead-acid batteries must always be kept in a fully charged condition. The terminal voltage can be measured as an indication of state of charge. Batteries may be charged periodically by a constant voltage method, or attached to a float charger.

Changing a battery
In modern automobiles, the grounding is provided by connecting the body of the car to the negative electrode of the battery, a system called negative ground. In the past some cars had positive ground. Such vehicles were found to suffer worse body corrosion and sometimes blocked radiators due to deposition of metal sludge.

When changing a battery, battery manufacturers recommend disconnecting the ground connection first to prevent accidental short-circuits between the battery terminal and the vehicle frame.

The majority of automotive lead-acid batteries are filled with the appropriate electrolyte solution at the manufacturing plant, and shipped to the retailers ready to sell. Decades ago, this was not the case. The retailer filled the battery, usually at the time of purchase, and charged the battery. This was a time-consuming and potentially dangerous process. Care had to be taken when filling the battery with acid, as acids are highly corrosive and can damage eyes, skin and mucous membranes.

Car batteries should be installed within one year of manufacture. The manufacturing date is normally printed on a sticker on the top of side of the battery.

When first installing a newly purchased battery a top up charge at a low rate with an external battery charger (available at auto parts stores) may maximise the battery life and minimise the load on the vehicle charging system. The top up charge is considered complete when the terminal voltage is just above 15.1 V DC. 15 V DC is the voltage level where any sulphation that may be present is driven from the plates back into the electrolyte solution. A new battery can have some sulphation even though it has never been in service. If the top up charge cannot be done it is not harmful to place the battery in immediate service.

Common battery faults include:

- shorted cell due to failure of the separator between the positive and negative plates
- shorted cell or cells due to build up of shed plate material building up below the plates of the cell
- broken internal connections due to corrosion
- broken plates due to vibration and corrosion
- low electrolyte
- cracked or broken case
- broken terminals

Sulphation - after prolonged disuse in a low or zero charged state corrosion at the battery terminals can prevent a car from starting, by adding electrical resistance. The white powder sometimes found around the battery terminals is usually lead sulphate which is toxic by inhalation, ingestion and skin contact. The corrosion is caused by an imperfect seal between the plastic battery case and lead battery post allowing sulphuric acid to attack the battery posts.

The corrosion process is also expedited by over-charging. Corrosion can also be caused by factors such as, salt water, dirt, heat and humidity in the air, a crack in the battery casing, or loose battery terminals. Inspection, cleaning and protection with a coating are measures used to prevent corrosion of battery terminals.

Sulphation occurs when a battery is not fully charged, and the longer it remains in a discharged state the harder it is to overcome the sulphation. This may be overcome with slow, low-current (trickle) charging. Sulphation is due to formation of large, non-conductive lead sulphate crystals on the plates; lead sulphate formation is part of each cycle, but in the discharged condition the crystals become large and block passage of current through the electrolyte. The primary wear-out mechanism is the shedding of active material from the battery plates, which accumulates at the bottom of the cells and which may eventually short-circuit the plates. Early automotive batteries could sometimes be repaired by dismantling and replacing damaged separators, plates, intercell connectors, and other repairs. Modern battery cases do not facilitate such repairs. An internal fault generally requires replacement of the entire unit.

Exploding batteries
Any lead-acid battery system when overcharged will produce hydrogen gas. If the rate of overcharge is small, the vents of each cell allow the dissipation of the gas. However, on severe overcharge, if ventilation is inadequate or the battery is faulty, a flammable concentration of hydrogen may remain in the cell or in the battery enclosure. Any spark can cause a hydrogen and oxygen explosion, which will damage the battery and its surroundings and which will also disperse acid into the surroundings. Anyone close to the battery may be injured. Sometimes the ends of a battery will be severely swollen, and when accompanied by the case being too hot to touch, this usually indicates a malfunction in the charging system of the car.

Reversing the positive and negative leads will damage the battery and may lead to gassing and explosion. When severely overcharged, a lead-acid battery gases at a high level and the venting system built into the battery cannot handle the high level of gas, so the pressure builds inside the battery, resulting in the swollen ends. An unregulated alternator can put out a high level of charge, and can quickly ruin a battery.

A swollen, hot battery is very dangerous, and should not be handled until it has been given sufficient time to cool and any hydrogen gas present to dissipate. A person handling a car battery should always wear proper protective equipment (goggles, overalls, gloves) and make certain there are no sparks or smoking close by.

Ampere-hours (Ah) is the product of the time that a battery can deliver a certain amount of current (in hours) times that current (in amperes), for a particular discharge period. This is one indication of the total amount of charge a battery is able to store and deliver at its rated voltage. This rating is rarely stated for automotive batteries, except in Europe where it is required by law.

Cranking amperes (CA), also sometimes referred to as marine cranking amperes (MCA), is the amount of current a battery can provide at 32 °F (0 °C). The rating is defined as the number of amperes a lead-acid battery at that temperature can deliver for thirty seconds and maintain at least 1.2 volts per cell (7.2 volts for a 12 volt battery).

Cold cranking amperes (CCA) is the amount of current a battery can provide at 0 °F (-18 °C). The rating is defined as the current a lead-acid battery at that temperature can deliver for thirty seconds and maintain at least 1.2 volts per cell (7.2 volts for a 12- volt battery). It is a more demanding test than those at higher temperatures.

Hot cranking amperes (HCA) is the amount of current a battery can provide at 80 °F (26.7 °C). The rating is defined as the current a lead-acid battery at that temperature can deliver for thirty seconds and maintain at least 1.2 volts per cell (7.2 volts for a 12- volt battery).

Reserve capacity minutes (RCM), also referred to as reserve capacity (RC), is a battery's ability to sustain a minimum stated electrical load; it is defined as the time (in minutes) that a lead-acid battery at 80 °F (27 °C) will continuously deliver 25 amperes before its voltage drops below 10.5 volts.

Battery Council International group size (BCI) specifies a battery's physical dimensions, such as length, width, and height. These groups are determined by the Battery Council International organization.

Peukert's Law expresses the fact that the capacity available from a battery varies according to how rapidly it is discharged. A battery discharged at high rate will give fewer ampere hours than one discharged more slowly.

The hydrometer measures the density, and therefore indirectly the amount of sulphuric acid in the electrolyte. A low reading means that sulphate is bound to the battery plates and that the battery is discharged. Upon recharge of the battery, the sulphate returns to the electrolyte.

Terminal voltage
The open circuit voltage is measured when the engine is off and no loads are connected. Open circuit voltage is also affected by temperature, and the specific gravity of the electrolyte at full charge.

The following is common for a six-cell automotive lead-acid battery at room temperature:

- Quiescent (open-circuit) voltage at full charge: 12.6 V
- Fully discharged: 11.8 V
- Charge with 13.2–14.4 V
- Gassing voltage: 14.4 V
- Continuous-preservation charge with max.13.2 V

After full charge the terminal voltage will drop quickly to 13.2 V and then slowly to 12.6 V. Open circuit voltage is measured twelve hours after charging to allow surface charge to dissipate and enable a more accurate reading. All voltages are at 20 °C, and must be adjusted -0.022V/°C for temperature changes.

GLOSSARY

ALTERNATOR
A belt driven motor on the engine that produces alternating current and converts it to direct current. The current is stored in a battery and used to supply all electrical needs of a vehicle.

ALTERNATOR BRUSH

ALTERNATOR PULLEY
A disk with single or multiple radial grooves that mounts on a generator or alternator shaft for transmitting movement.

AMPERES (amp)
An electrical unit of measure used to indicate alternator output at a specified battery voltage.

ARMATURE
The rotating part of a DC motor with a commutator.

B TERMINAL
A terminal connected to the battery positive.

BATTERY

An electrical storage device which is used to hold electrical current for use in starting a vehicle and operating certain equipment during periods of high load.

BENDIX
An inertia-type starter engagement mechanism.

BRUSH
A device which makes the electrical contact to the commutator on the armature of a DC generator or motor, or to the slip rings on the rotor of an alternator.

C TERMINAL
A circuit used to interface with the vehicle computer through the alternator control unit. The computer regulates through this circuit.

CAPACITOR
A device that stores charge like electrons or the absence of electrons(positive charge).

CLUTCH PULLEY
Overrunning Alternator Pulley

COMPUTER
An on board computer which controls fuel mixture, ignition timing, and in some vehicles, the
voltage regulator, plus other functions.

D TERMINAL
A dummy terminal

DE FRAME
Front section of an alternator case.

DIODE
Diodes are located in the rectifier and operate like an electrical gate allowing only positive or negative voltage to pass through the rectifier. This changes the AC voltage generated by an alternator to DC voltage which can be used in a vehicle's DC electrical system.

DIODE TRIO
A separate mini rectifier which supplies DC voltage to the regulator to energise the alternator field. Most IC alternators and some EXT alternators contain diodes trios to operate the regulator.

DIRECT DRIVE STARTER
Typically a light duty or older type of starter in which the drive gear is attached directly to the armature shaft and rotates at armature speed during operation.

F Terminal
Field and Feedback Terminal

FIELD
This is the electrical field generated in an alternator by the rotation of the rotor inside the stator coils. Many EXT alternators have a terminal marked with F to indicate the field connection.

GEAR REDUCTION STARTER
A starter which contains additional drive gears between the armature shaft and final drive gears to provide greater power to start the engine. These units are typically found in later model applications.

STARTER MOTORS & ALTERNATORS

IAR REGULATOR
Internal Alternator Regulation Regulators found within an alternator.

IC
Acronym for Integrated Circuit. A device consisting of a number of connected circuit elements such as transistors and resistors, fabricated on a single chip of silicon crystal or other semiconductor material.

INSULATOR
A substance with very high resistance that conducts electricity poorly.

MAGNET
An object that is surrounded by a magnetic field and that has the property.

P TERMINAL
A separate low voltage output terminal on the alternator which is used to operate a relay. Typically operates a fuel pump or automatic choke.

R TERMINAL
A separate solenoid connection on some starters which allow battery current from the starter solenoid to operate a cold start relay.

RECTIFIER
An electronic device that converts alternating current into direct current.

SERPENTINE PULLEY
Late style of pulley which allows use of a flat drive belt. This design provides longer belt life than conventional V-belt designs.

SLIP RING
The part of an alternator rotor that is used to allow current to pass while it rotates.

SOFT START
Controls the field circuit to prevent overload at low RPM's.

SOLENOID
An electrical relay which uses electro-magnetic coils and a plunger to extend the drive gear into the flywheel and make contact between the battery and starter armature to turn the engine during starting.

SPRAG
A device that is placed between two cylinders that will allow the cylinders to lock up when one cylinder is rotated in one direction and released when it is rotated in the opposite direction.

STARTER
An electrical motor and drive mechanism designed to start an engine. The starter turns the crankshaft through a pair of gears.

STARTER DRIVE
The starter mechanism that is activated by the solenoid to drive the flywheel and turn the engine during starting.

STATOR
The non-moving part of an AC alternator that produces current.

STARTER MOTORS & ALTERNATORS

Typical alternator stators have three separate windings connected as Delta or WYE. A rotating rotor provides a moving magnetic field and induces a current in the winding of the stator.

VOLTAGE DROP
A common problem which normally occurs at several points in the electrical systems. It is most severe between battery and starter solenoid connections and commonly caused by corroded or loose wiring.

VOLTAGE REGULATOR
An electronic or electro-mechanical device used to control the voltage of a generator or alternator by controlling the field of the generator and the rotor current in an alternator.

VOLTS
An electrical unit of measure used to determine the operating voltage of a vehicle electrical system and battery condition. Volts are not used to indicate alternators output.

W TERMINAL
A separate low voltage output terminal on the alternator which is used to operate a tachometer or RPM warning light.

www.ingramcontent.com/pod-product-compliance
Lightning Source LLC
LaVergne TN
LVHW041633070426
835507LV00008B/588